RAISE THE ZONE

How to Light Up Earth's Luminous Age

JEFF PECKMAN

POWER UP Books

Publisher:
POWER UP Books, LLC
Littleton, CO
www.powerupbooks.com
info@powerupbooks.com

Peckman, Jeff
RAISE the ZONE: How to Light Up Earth's Luminous Age / Jeff Peckman
Includes index.

ISBN 978-0-9989092-3-3 (paperback)
ISBN 978-0-9989092-8-8 (ebook)

Library of Congress Control Number: 2026935694

1. Air pollution. 2. Technical innovations. 3. Social change. 4. Practical politics.
I. Title: RAISE the ZONE. II. Title.

Cover design by Jeff Peckman, except for the NASA photo of Earth, and Rainbow Light Ship by Chuck Chroma

Revised Edition

10 9 8 7 6 5 4 3 2 1
Printed in the United States of America

Dedication

To my beloved parents and "co-authors,"
Victor and Lorene, whose love, nurturing, patience, tolerance, and wonderfully precise guidance, made *RAISE the ZONE* possible, and made my life amazing!

Thank you!

Testimonials for the Book and Author

"*RAISE the ZONE* is a message of hope. It's a message of practical application. It's a message that will quench the thirst and hunger in people, right now, for a better society, and no longer a need to look elsewhere, but to look inside."

- Cam Vuksinich, Life Coach, Author,
Founder of One World Heart Project

Jeff Peckman has a natural talent and comfort level for dealing with the messy stages of a new project. He has been very quick to scope out the needs of the situation and organize the effort around available resources. Jeff is a natural leader and just has a real knack for getting things done."

– Thomas Frey, Executive Director of the DaVinci Institute

"In person, Peckman looks and sounds like your favorite college professor, the one who could get the whole class debating novel solutions to the world's problems."

– Lisa Ryckman, Rocky Mountain News, June 2008, currently Associate Director of Communications at the National Conference of State Legislatures

"Everyone's talking peace: Jeff Peckman has not put a single penny into advertising for his Safety through Peace initiative. He hasn't had to. The proposal, formally titled Initiative 101, has not only been the talk of the town but the talk of the nation." – Denver Daily News, 2003

"At the forum Tuesday, sponsored by environmental groups and focusing on conservation and sustainability, many of the candidates referred to Peckman's intelligence, some even asking to give their answers to questions before he did so as not to look less thoughtful by comparison when going after."

- Fox 31 KDVR.com, April 2011

Contents

Introduction **ix**

1: Journey of a Solutionist **1**

From a Hydrogen Bomb to a Kansas Farm *1*
Vowing to Solve Global Problems at Age 10 *2*
"You Should Be Called Blue Cloud" *6*
Four Feet of Hail Woke Me Up *8*
Air Pollution Solution for the Rich and Famous *8*
Robert Redford's Reply *9*

2: Tracking Truth in a Fake News Forest **14**

Truth Alone Triumphs *14*
Sharpening Occam's Razor *15*
Science of Lying *20*
Diagnosing the Fake News Epidemic *21*
Lying as a Weapon of Mass Destruction *23*
A Cure for Misinformation *25*
Is Truth Compatible with Freedom of Speech? *26*

3: Light Up the People **29**

Engaging the Power of the People *31*
"Plan B" from America's Founders *33*
Truth in Labeling GMOs *36*
Paul McCartney's Endorsement to "Let it Be" Labeled *38*
Safety through Peace – Initiative 101 & Counterterrorism *39*
Politicians Go "Ape Snark" *45*
Safety through Peace Makes Headlines *46*
Warned About the Conservative Radio Monster *47*

4: Finding Your Moonshot **50**

Snap Out of It – Expand Your Mind 50
Why America's Founders Would Enjoy Star Trek 51
Who Gains from Hiding Extraterrestrial Technology? 54
World Leaders' Views of Galactic Neighbors 57
Giving NASA and the Pentagon a Cosmic Nudge 58
Star Trek Founder's Fear and Warning 59
Benefits of Looking Up 59

5: Vote to Think Big **62**

Extraterrestrial Affairs Commission Campaign 62
Surprising Opinion Polls on Space Visitors 65
Extraterrestrial Visitors in the Media Spotlight 68
A "Street Level" View of Space Visitors 70
Take Us to Your Civic Leaders 71
Solving an Election Day Mystery 72
Views from Ethnic Groups About Extraterrestrial Visitors 74
Extraterrestrial Mega-Concerts Zapped by Greed 77

6: Roots of Our Problem-Based Economy **80**

Achilles' Heel of Democracy 82
Who Rules the Rulers? 82
Why Politicians Dismiss Citizens' Solutions 84
Government Ineffectiveness Explained 85
When Atomic Fracking "Bombed'" 90
From "Revolving Door" to "Evolving Door" 91
Virus of Incoherence 96
The Onus of Self-Government is On Us! 97

7: Elevate Your State – Be a Country! **98**

Rising Above the Duopoly 98
What Would the Founders Do? 99
Who's a Terrorist? 101
America's Epic Predicament 102
Declaring Independence Again 105
Ascension Instead of Secession 108
Manifesting a Higher Manifest Destiny 111
Raise the Body Politic 114

12: Global LOVE and Coherence **184**

A Simple Do-It-Yourself Approach 184
Plague of Measurement Uncertainty 185
Coherent Electromagnetic Fields 186
Rise of COCUUN™ 186
Coherent Fuel Plasma 188
Vehicles as Air Purifiers 189
Coherent Water 189
Creating a Global Coherence Grid 190
Music for Healing and Balance in Nature 191
Creating a LOVE-Based Economy 192
Light Up a Global Transformation 195

13: Push-Button Air Pollution Solution **197**

Lift Air Quality with an Archimedes "Lever" 197
Allowing Children to Breathe 198
How "Greased Lightning" Could Clean Our Air 200
Shortcut to Reduce Greenhouse Gases 200
Global Outdoor Air Purifier 202
Global Clean Air "Earthshot" 204
Getting Big Drivers in the Race 205
We are the "Great Leader" 207
The People's "First 100 Days" 208
Call to Action and a Word of Caution 209

"End of Tour" Souvenir **210**
Acknowledgements **211**
Glossary **212**
Appendix A: The People's "First 100 Days" **213**
Appendix B: Ballot Initiative Concepts **215**
Index **219**
Ending at the Beginning **223**

8: Harnessing the Technology Supernova **117**

Political Injustice for Criminal Justice *118*
Greenwashing in Politics *119*
Does Technology Suppression Benefit the People? *122*
Is the Patent Office Patently Unfair? *126*
Protecting Extraordinary Technologies *127*
How is Your "Vision"? *130*
Growing a "Y.E.S." Culture of Solutions *130*
Politicians are the Last to Know *132*

9: Light Up a Solution-Based Civilization **135**

Conscious Capitalism *136*
Enlightened Self Interest vs. Selfishness *137*
Flawed Notions in Free Market Philosophy *137*
Are Selfishness and Altruism Opposites? *138*
"Enlightened" is the Key *143*
Sliding Scale from Selfish to Selfless *145*
Cosmic Self-Identity *146*
Expanding Selfness of the Masses *148*

10: Truth, Coherence, LOVE – Antidotes to Stress **150**

Harmony of Truth, Coherence and LOVE *151*
Using Harmony to Get Unstuck *153*
Intelligence of the Heart *156*
How Stress Amplifies Weaknesses *157*
Stress Triggers Unnecessary Genetic Expression *158*
Stress Causes "Swiss Cheese" Brains *160*
Collective Stress Affects Communities *160*
Rishis, Saints and Scientists *164*
Antidotes for Global Stress *165*

11: Giving Xperiment **168**

Better Health in 15 Seconds *168*
Unlimited LOVE and Instant Health *169*
Giving Xperiment Discovers "Amazing Mouse" *171*
Butterflies Surfing LOVE Waves *178*
Moon-walker Mitchell Finds Cosmic Connection *181*

Introduction

RAISE the ZONE is about raising the coherence of individual and collective consciousness to light up Earth's luminous age. That is achieved through truth, LOVE, and coherence. It's part memoir, and mostly a blueprint of how to solve problems better on a personal, societal, and global scale, affordably and quickly. *RAISE the ZONE* reveals the root cause of persistent problems. It offers practical multi-partisan solutions to inoculate humans from the virus of incoherence that plagues humanity.

RAISE the ZONE is opposite of "flood the zone" which is all about lies, chaos, and incoherence. I'm your tour guide. Along the tour, I'll share the arc of my journey from a dark view of the political landscape to "seeing the light" and a glowing path toward a better world. Laughing at my occasional tour guide jokes is optional, but encouraged. Are you ready a new kind of adventure?

On the book cover, the color blue is a universal symbol for truth, as in "true blue". "Light Up" is the activation. "Luminous" is the continuation of Earth illuminated with the light of universal LOVE, truth, coherence, and all that radiates the highest and best of humanity. Illumination is a light-filled nation.

Vesica Pisces circles, on the cover and chapter titles, are a concept from Sacred Geometry. In *RAISE the ZONE,* they represent opposite, or conflicting, interests at the root of all problems. The center represents balancing these opposites at the unifying point of universal LOVE, from which the *fabric* of creation emerges as the vast and expanding universe. The starry sky looks like fabric because it's actually a photo of my blue sweater with white specks!

LOVE is the essential and overarching theme that connects every part of this tour. LOVE *is* the ZONE! LOVE is in ALL CAPS to mean universal and unlimited LOVE, which is different from romantic love that is not our primary focus on this tour. LOVE harmonizes everyone and everything at the deepest level.

RAISE the ZONE Tour Map

We'll explore truth, fake news, and brain waves. We'll dig for causes of major problems on Earth, and also reflect on extraterrestrial life. We'll examine decades-long obstructions of novel solutions to major problems of all kinds. Earth's *luminous age* is not about gold and opulence for the super-wealthy. It's about precious opportunities and values of an enlightened civilization.

The middle chapters introduce coherence technologies that can transform our world from a problem-based economy into an enlightened, solution-based civilization. The later chapters focus on strategies for raising the zone of humanity on a global scale.

The final chapter offers an enlightened, cost-effective solution for solving global air pollution with the "push of a button." It will propose a moonshot, or rather an *Earthshot* project, which could have a domino effect of benefits worldwide.

Accelerating Our Planetary Shift

An increasing number of people worldwide are feeling their true calling. They may face doubts, ridicule, hate, and resistance to change, but they're raising Earth toward a brightly lit paradigm. These activists, entrepreneurs, visionaries, inventors, civic leaders, discoverers, and adventurers are in every field of life and every country. Something lit a fire under them. Each has their spark. The radiant "stars" on the Earth image honor their service.

Is solving human-made climate change, or at least global air pollution, your highest priority? If so, rush to the last chapter right now! Otherwise, take the leisurely, well-lit scenic tour.

You'll learn that achieving an upgrade for humanity is easier and more affordable than most people know. Best of all, it doesn't take government action to accomplish it. We can do it ourselves as ordinary citizens. We must do it because too many politicians are inherently unwilling and incapable of preventing and solving problems that we can solve ourselves.

RAISE the ZONE explains why you have more authority and freedom than any politician or government. *You* rule the rulers. *RAISE the ZONE* can empower you to be your best guide.

Raise the Zone Exponentially through Togetherness

There is strength in numbers. There's exponentially greater strength in loving and coherent numbers working in harmony. Even a very small number of committed and persistent people can create a world-changing paradigm shift.

Disclosure

Unless otherwise noted, the opinions and observations I've shared are my own. They do not necessarily represent the views of anyone else who assisted in the production of this book.

A Call to Action

RAISE the ZONE is an urgent call to action. It's a command to fully express the highest and best of yourself and humanity. The application any one of the solutions in *RAISE the ZONE* could be life-changing and even world-changing. So, raise the zone, light up your life from the inside, and radiate your star light to others!

Remember to look for the LOVE that connects everything!

Upward and Onward!
Jeff Peckman
RAISE the ZONE Tour Guide

1
Journey of a Solutionist

From a Hydrogen Bomb to a Kansas Farm

Are you ready to *raise the zone* and create a solution-based world? Our starting point is where my journey, or my tour of life, began. On March 1, 1954, I was born in Paola, Kansas, about 17 hours after the Castle Bravo nuclear blast lit up the South Pacific. It was the largest explosion of a hydrogen bomb by the U.S. in history – 1,000 times greater than the bomb dropped on Hiroshima. It was 2.5 times larger than expected, due to a miscalculation.

Above is a photo of my late father's painting of our family farm about one mile outside of Paola. It was a wonderland of learning and adventure. It also taught me a little about nature's mysteries and magic. Five years on our farm set me on a quest for truth and prepared my life's mission as a "solutionist".

Are *you* a solutionist? That is someone who solves puzzles or problems on any scale. It best describes my life and mission. I would add *discoverer of possibilities* to that definition. Activists and advocates do vital work in raising awareness concerning a range of problems. My approach is to offer proven and affordable solutions. This chapter shares my journey that led me to write a book about uplifting your life and our world.

Vowing to Solve Global Problems at Age 10

My path of solutionism began in elementary school with a big wake-up call. It happened on a playground – not in class. Kids talked about all kinds of things. One day, a friend walked up to a few of us who were in a conversation. He claimed, "My dad said you're using only one tenth of your brain." That ignited my anger and sparked my curiosity. It probably had roots in American psychologist William James' view that people rarely achieve more than a small percentage of their full human potential.

Regardless of the validity of that view, it shook me to my core and infuriated me. After my shock settled down, it hit me that if his claim was true, it might explain some terrible things. Those included JFK's assassination a year before, the Cold War, escalation of the Vietnam War, and maybe even bullying at school.

My rude awakening compelled me to vow two things. First, to develop my full brain. Second, to find solutions for big and little problems in our world and life in general. It seemed simple. Why would a 10-year-old child dare to have such bold goals? I didn't ask. My path and vision were clear and illuminated. No reason came to mind why these goals were not achievable.

How does one start down those paths? I did what everyone does who has ever achieved anything – I had an intention to do something and committed to it. Those are common steps advised by probably every motivational philosophy. Major marketing campaigns are also built around those same essential principles to promote goods, services, institutions and more.

Starting on My Paths

Those first steps on my paths were challenging. There were no maps or signs, but resources quickly came to my attention. Plato's Republic was one. Other comments from unexpected sources helped too. Balance in everything seemed important. Mind-body coordination and eating right were part of that.

About that same time in elementary school, I got straight A's and won the most "Blue Ribbons" on field day. I taught myself to do handstands and handsprings. A year later I was elected student council president. However, it became apparent that I could do little to help anyone in my school. It didn't propel me into politics. Instead, it inspired me to find *real* solutions in other ways.

Crystallizing a Better Future

Another story in elementary school greatly influenced my life. A friend told me about an unusual group of people. They would find out when lakes or ponds were nearing the point of freezing. They would carefully sneak up to a body of water. With the most delicate touch, one person would place a seed ice crystal into still water. Suddenly, as they expected, the water's surface started to crystallize. It made a unique crackling sound. The delicate patterns spread in all directions. It must have been thrilling to see and hear that magical phenomenon in nature.

Was that event possible? I didn't know, but I have seen many opportunities throughout life to inject a tiny crystal of an idea into situations at what turned out to be perfect timing. I believe that such opportunities exist right now around our world. Each could allow huge advances for humanity with the right "crystal" idea. What's your crystal idea?

Coherent symmetrical designs, like those of crystals, always fascinated me. Following, are scans of some of my drawings from elementary school. A piece of paper, a pin to poke holes, a ruler and some colored pens, were all I used. I believe these images planted seed crystals for writing this book.

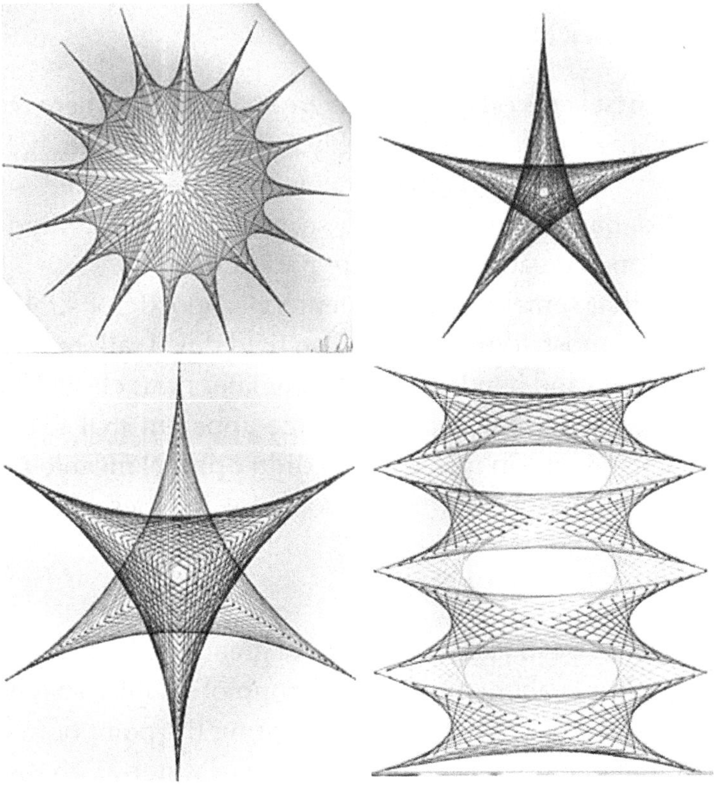

Choking on Rocks to Speak Better

I read about Demosthenes, the prominent Greek citizen and orator in ancient Athens. He put a pebble in his mouth and spoke out over the sea to improve his speech. I talked with small objects in my mouth over the parking lot next to our house. Alas, after swallowing or choking on too many rocks, marbles and quarters, I moved on to other ways to develop my full brain.

My attempts included trying a more balanced and healthy diet. That meant primarily one thing to me. I would take some of just about everything in the refrigerator and pile it all between two slices of white bread. The sandwich would be about three inches high. After eating it, I would wash it down with gulps of milk, then carbonated soda, orange juice and Kool-Aid. I drank straight from all the containers. Life seemed balanced! I'm not sure my stomach was. At least my intentions were good. A few years later, I learned about truly healthy habits such as exercise, cleansing my lymphatic system, and growing and eating organic vegetables.

Meditating to Survive Teenage Life

In junior high school, I tried juggling, judo and yoga. They all felt right and good. I also got into mild drugs. That was not good, but I gave into peer pressure and curiosity. In high school, I went to a lecture on *Transcendental Meditation™* (TM). The TM teacher said, "Psychologists tell us we use only 5-10% of our mental potential. TM helps to develop the rest of our brain."

That convinced me to learn TM right away. In my first experience of TM, a huge cloud of stress lifted from me. The natural high that I felt from TM was so much better than cannabis or other drugs, that I immediately lost interest in them. That was my first real enlightening experience on the path to develop my full human potential. I've continued my daily practice of TM since then.

Buckminster Fuller's Influence

Buckminster Fuller was another great inspiration. The story I remember about Bucky Fuller is that he was despondent about his young daughter's death and considered ending his life in Lake Michigan. As he stood there in deep water and deeper sorrow, contemplating the last moments of his life, he pondered a big question, "How much can one person achieve?" His curiosity compelled him to walk out of the lake.

Fuller went on to become a legendary inventor, futurist, and visionary. He inspired millions with his futuristic, and yet, practical solutions. His "Dymaxion Car", geodesic domes, floating cities, and "Spaceship Earth" philosophy, were legendary.

His extensive research and engineering concluded that it was possible to make war obsolete and provide everyone on Earth with an extremely high quality of life by converting the world's deadly "weaponry" to "livingry".

"You Should Be Called Blue Cloud"

Vivid dreams have been common for me since childhood. Some were lucid, in that I seemed to be in control. I could explain things during a dream as clearly as if in waking state. A few dreams have shown me things that did not exist at the time or had not yet happened. They then occurred later in remarkable matching details.

One particular dream, in 2009, was both symbolic and prophetic. It triggered my renewed promotion of one of the most extraordinary solutions I know. It was powerful in its impact and played a big part in inspiring me to write this book.

In the dream, I was on a public bus going north on Federal Boulevard in Denver. I've never taken that route on a bus. The only people I saw on the bus were three Native American children on seats across the aisle and me. They looked at me curiously for a short time. One of them asked my name. I answered "Jeffrey." They looked at each other in a disapproving way. Then one of them said, "That is not a very good name for you. You should be called Blue Cloud." That simple response surprised me. Why would they call me that? I had never heard the name of Blue Cloud before. We then all sat quietly without talking.

Our bus stopped one block before the major intersection of West Alameda Avenue and Federal Boulevard. It's next to what is now the Far East Asian shopping center. We all got off the bus. They knew where to go. I followed them out of curiosity. We walked west along a side street in an area that my family drove through every Sunday to a Lutheran church. It was three blocks west of where we got off the bus.

We arrived where the church would have been. However, the street intersection, and nearby area, was instead a pristine, grassy, open clearing surrounded by pine trees, as I would expect to see in the Black Hills of South Dakota. In the clearing were four Native American adults. Two of them were 30-40 feet in front of me. Two were about twenty feet to my left.

Close in front of me was a row of at least twelve colorful, closed, egg-shaped baskets. They had traditional Native American designs on what looked like large Easter eggs. Colors were vibrant and their designs were intricate. The baskets laid on top of wooden tables about four feet high. It was a wonderful showcase of Native American artistry. There was no verbal conversation with those adults. Then, through what seemed to be both intuition and clear telepathic communication, my heart sank in deep sadness.

The row of baskets was a group of papooses. One definition of papoose is a backpack or basket to carry a Native American child. Another definition of papoose is the child itself. The raised tables, or platforms, were for cremation of the children. Telepathically, the adults told me that the infants had died from a "harsh winter". Grief overwhelmed me, and I could only weep from that image. In our present time, we think of harsh winters and extreme weather events from human-caused climate change.

Vision of Preventable Disaster

Seeing my deep sadness, they quickly explained that it was not a vision of only one possibility. They said the harsh winter had not happened yet and it did not *need* to happen. They said there were things I could do to prevent it. Naturally, I was relieved and yet confused. What could I do? What could anyone else do? I woke up, literally and metaphorically. Immediately, I went to my desk and searched the Internet for "Blue Cloud". I found a few references. One said it was a cloud containing nourishing rain. To me, that meant balance in nature and the good fortune of rain.

That dream strongly affected my plans going forward. I looked for clues about what I could do to prevent the harsh winter. How could I help avoid the tragic deaths of such precious children in those beautiful baskets? Was it a premonition about children all over our world? Months later, a detail finally struck me.

We got off the bus at South Federal Boulevard and West Dakota Avenue. Symbolically, I had gotten off the government

(Federal) bus and walked with the Native American (Dakota) children in the pristine environment that they regard as sacred. I kept searching for any means to prevent that harsh winter, concerned that it was not just a dozen children, but also much of human life.

Four Feet of Hail Woke Me Up

Jumping to June 2015, another shockwave hit me. This time, it was in my waking state of real life. A major hailstorm hit Denver on June 4. It happened to be "bike-to-work" day. I was at home, frantically protecting my amazing organic strawberries just as the intense storm began. The next day, the *Denver Post* reported:

"One Denver block buried under up to 4 feet of hail."

The article stated that the affected area was between Dakota Avenue and Alaska Street. I was awestruck. That area along South Irving Street was exactly where I stood with the Native American adults and children in my dream six years earlier. The message was clear. It was time to speed up promoting a breakthrough global solution to climate change.

An Air Pollution Solution for the Rich and Famous

It was still June. Many top "A-list" celebrities had made headlines for supporting different campaigns about air pollution and potentially catastrophic effects of climate change. It was time to inform them of a unique technology that had reduced Los Angeles air pollution at least 40% over a three-month period in 1990!

One month after the rare hail pile-up in Denver, I sent a letter to over 50 of these A-list stars. It was a long shot, as you never know if your letter will get past the filter of agents, publicists, and other gatekeepers. Only about five of the letters came back to me. However, no one replied. It didn't matter by that time because my priorities suddenly shifted.

My dear mother, whom I was caring for, was experiencing worse respiratory health problems. That's a common effect of high levels of ozone. However, all I could do was to make her as comfortable as possible in her remaining days. She finally passed away that summer in August. It reinforced my commitment to share a solution to air pollution and its related health disorders.

Robert Redford's Reply

In September 2015, I received a kind and appreciative letter from Mr. Redford's assistant. He said that Redford had too much on his plate to get involved in what I proposed. So, it was clear that I should do myself, what my proposal asked of celebrities.

In October 2015, I wanted to replicate in Denver, the L.A. air quality study done by the inventor in 1990. However, he wanted to finish some scientific explanations for his technology. So, it seemed like a good time for me to write my first non-fiction book: *POWER UP – How to Create a Solution-Based World in 100 Days.*

Now, with revised content and new name of *RAISE the ZONE,* I'll add my favorite full quotation about commitment. That's a big theme throughout our tour. The following quotation is from the famous explorer and Mt. Everest climber, William H. Murray:

> This may sound too simple but is great in consequence. Until one is committed, there is hesitancy, the chance to draw back, always ineffectiveness. Concerning all acts of initiative (and creation), there is one elementary truth, the ignorance of which kills countless ideas and splendid plans: that the moment one definitely commits oneself, then providence moves too. A whole stream of events issues from the decision, raising in one's favor all manner of unforeseen incidents, meetings and material assistance, which no man could have dreamt would have come his way. I learned a deep respect for one of Goethe's couplets:
>
> Whatever you can do or dream, you can begin it.
> Boldness has genius, power and magic in it!

The Magic of Commitment – A Surreal Story

As a warm-up lap before our main tour starts, here's one story that illustrates the magical power of commitment. It also shows one of the first times when backroom politics directly influenced my life. During high school, I envisioned my future as an architect. I didn't go immediately into college for that, but instead became a teacher of Transcendental Meditation. Still, solar architecture and urban planning remained a strong interest.

Then, in 1979, I unexpectedly met Colorado's renowned solar architect, Richard Crowther, at his new passive solar home. We introduced ourselves and shared our interests. He wanted to learn about Transcendental Meditation. I wanted to learn about architecture. Subsequently, we made an agreement that he would take me on as an apprentice and assistant property manager.

My wife at that time, and I, were to live in one of the units of his new home and receive a good salary. It was a dream job. However, before we moved in, we were called to attend a World Peace Assembly due to the tense relations between the U.S. and U.S.S.R. The situation was dire enough, that all of the conference attendees were urged to continue in a program for many more months. So, I had to withdraw from the agreement.

That did not end my interest in architecture. In 1995, I worked at a construction company that was introducing *Vastu Shastra,* also known as *Sthapatya Veda,* into building and town design. It's the ancient science of building in accord with natural law. Then, my Italian friend in Rome asked me if I could organize a team for a "by-invitation-only" international architectural design competition. I said, "Sure! I can do that". I committed.

I knew a few contacts from my job. The design competition was for redeveloping the Piazza dei Navigatori, in the EUR, south of Rome on Via Christoforo Columbo. The subject six-acre property was owned by my friend and his stepbrother, a construction engineer. It was the largest undeveloped site in Rome. It had been mostly vacant since World War II, when their father owned it.

In February 1999, my friend paid for me to visit Rome for some meetings with his stepbrother and team. After arriving in the morning from an overnight flight to Rome, no sleep, and no place to stay yet, he said we needed to attend a pre-wedding party for his cousin. We arrived about 9:30 p.m. in Bracciano. The party was in the famous Odescalchi-Orsini Castle on top of a hill overlooking Lake Bracciano. The high-society event was for his cousin and bride-to-be, who owned the castle. I quickly felt more out of my element than I had ever felt before. However, the guests I met were wonderfully gracious and personable.

European jetsetters were arriving as the older wealthy guests were leaving. One of my conversations was with Mark, an American friend of my Italian friend. I told him about my reason for coming to Italy. He told me about his new photography business and non-profit work that he and his wife did in Africa. I encouraged him to support a new school in South Africa called CIDA City Campus, which intended to provide nearly free university education to very economically disadvantaged South Africans.

At the end of our chat, he gave me his business card. It was Mark Getty, owner of *Getty Images* – still in its early years. Later, I learned that the Getty Foundation had made a large donation to CIDA. Moreover, the school became so successful and respected worldwide, that Sir Richard Branson, Oprah Winfrey, and other philanthropists gave large donations to CIDA. Taddy Blecher, the CIDA headmaster, achieved global recognition and praise.

Another conversation at the party was with an art dealer who asked about my visit. I said that my potential list of team members included the town planning firm that designed Seaside, Florida, the filming location of the Truman Show. I had not talked with them yet, though. He threw his hands up and dramatically burst out, "OH! The Truman Show. The TRUMAN SHOW! Is a Classique. A CLASSIQUE! Everybody in Italy love the Truman Show!" When Tom Cruise and Katie Holmes were married in the same castle in 2006, I reflected on my surreal experience at the pre-wedding party of the castle owners.

Back to the design competition. My friend's stepbrother had doubts about my value to the project. He knew I was obviously not an architect and quickly tried to discourage me. He firmly stated that we couldn't win. He said the winner had to be a team from Rome for political reasons. He also said that five teams had already been accepted, and only five were allowed.

I didn't know if he was stating facts, or speculation. I replied that it was supposed to be an international competition, but the five accepted teams were all European. I said it should have a U.S. team. Second, I added, that "it doesn't matter if we can't win, we'll still have the best design. In fact, it will be so good that people will wonder why we *didn't* win. It might even cause a scandal!" He was clearly in disbelief at my audacity. So, I added, "Give me two weeks. If I can't put together a team that's impressive enough to get accepted, I'll go away." He agreed. We had a deal.

Once back to the U.S., I quickly contacted two architects I knew who had impressive credentials. Neither of them took interest. That surprised and disappointed me. My last resort was the firm of Duany Plater-Zyberk that designed Seaside, Florida. I didn't know them personally, so just made a cold call and got a message machine. It was a long shot. Then the magic happened.

Jeff Speck, one of the town planners at DPZ, excitedly called me back. He was interested. A map of Rome filled an entire wall of his apartment. Later, I also found out that Andres Duany and Elizabeth Plater-Zyberk had previously lived for two years in the neighborhood of my friend's property!

The competition organizers accepted the DPZ team and made one of the other teams go away. In January 2000, I met with DPZ for a five-day design "charette" in Miami. On day one, it was obvious that my concept for applying Vedic design principles clashed with 'New Urbanism' concepts pioneered by DPZ. Some tension filled our small conference room. I took a break for my daily meditation. On the way back to the design room, I randomly grabbed a book on landscape architecture from DPZ's extensive library. One photo of a fountain closely resembled a Vedic design.

The team agreed to try the symmetrical fountain layout as a template for the overall design. To our surprise, my Italian friend flew in from Italy to check on our progress. He was intrigued by our design concept drafts. He asked, "where did you get this?"

I handed him the open book and said, "it was from this place called Villa Lante". He burst out laughing hard as he doubled over. We all recoiled – wondering if he thought we made some horrible design mistake. When he stopped laughing, he said with a big grin, "my mother was born there! My grandfather owned it!" Our great relief triggered a flash flood of exciting ideas. Mutual respect and harmony surged within our group.

The end result was a spectacular design by our stellar team of top architects. Even I contributed some good ideas that came from pure inspiration. When the juried competition was held in May of 2000, Jeff Speck reported that three of the celebrity architectural firms were dismissed within a few minutes. He said our design was the most "idea rich". After that, a contentious meeting behind closed doors, and a single extra vote, ultimately decided in favor of the Rome design team. My prediction came true.

A jury member suggested that we could get an "Honorable Mention". However, that doesn't really happen in such competitions. It didn't matter, either. The whole experience was fantastic and surreal from beginning to end. It was a perfect example of "a whole stream of events...raising in one's favor all manner of unforeseen incidents, meetings and material assistance, which no man could have dreamt would have come his way."

Summary

My journey as a solutionist has led me to many surprising places, unusual situations, and amazing inventions and inventors that I never dreamed were possible. *RAISE the ZONE* shares some of those stories to illustrate the power of making a commitment to achieve any goal, no matter how big or seemingly impossible. Now, let's get this tour rolling.

2

Tracking Truth in a Fake News Forest

"It is easy to accept a lie that supports what you want to hear."

Truth Alone Triumphs

Let's move our tour into the Hall of Truth, *if* we can find it. Feel your way along walls of an open and discerning mind. Are you ready to sharpen your truth-knowing skills? Any journey requires discovering truth at each step. Which is the correct path? How does one know? What signposts are along the way? Are they real? Are they illusions or merely dreaming? Are they tricks by selfish people wanting to exploit you?

Discovering the truth in any area helps humanity to advance, even if in small ways. Would you like to see more truth in our world? You can know truth through intellectual analysis. You can also know it through your intuition and "gut feeling." Pursuing truth always reveals surprises. Speaking truth always gets results.

An ancient Sanskrit phrase is "Truth Along Triumphs." Some truths are "self-evident" as America's founders stated: "...that all men are created equal, that they are endowed by their Creator with certain unalienable Rights, that among these are Life, Liberty and the pursuit of Happiness." It seems that most people have some truth to share that will benefit others.

Sharpening Occam's Razor

On the intellectual side, something called "Occam's razor" is a useful tool in pursuing scientific or philosophical truth. William of Occam, or Ockham, was a Franciscan friar who lived around 1285-1347 A.D. He developed this tool of analysis, according to many historians. The first time I heard of Occam's razor was in the movie *Contact*. The theme of both the book and film was a possible first encounter with an extraterrestrial civilization.

When watching *Contact*, the topic of Occam's razor went right by me, as did other scientific jargon. I never intended to write about it until now. Then I started hearing about Occam's razor again, but in debates over *real* UFO sightings. These were on TV talk shows by Larry King and others. People would often define Occam's razor as, "the simplest explanation tends to be the right one." What puzzled me was that both sides used it in debates.

Skeptics and debunkers would say something like, "Occam's razor—the simplest explanation is the right one." Then they would claim it's impossible for any extraterrestrial craft from the nearest inhabitable planet to reach Earth. Therefore, it has to be something else such as ball lightning, swamp gas, a top-secret project, natural phenomenon, and so on.

In one case, a highly trained Air Force officer spoke of witnessing a craft that he thought was obviously not from Earth. In response, a debunker casually said that the officer "misinterpreted his experience", but the debunker didn't provide any evidence or logic to support his claim.

Then there are strong believers in UFOs and extraterrestrial visitors who would say things like, "Occam's razor—the simplest explanation is the right one." Then, they declare that there must be other beings in outer space, so, it's obvious that any strange phenomenon in the sky is an extraterrestrial craft. Surprisingly, each side would use Occam's razor as a shield to defend their beliefs and dismiss opposing views. Does that make any sense? How could two opposing sides use the same respected truth-finding tool and come to such opposite conclusions? Rather than remain confused by this apparent contradiction, I did some research.

Firstly, Occam was a monk who probably did not even *own* a razor. I thought there must be an explanation and found out that Occam's razor was being misquoted. Both the book and movie *Contact* quoted it incorrectly, and did so in ways that were different from each other.

Occam's razor is meant to narrow down which hypothesis to study first. When there are multiple potential explanations for a phenomenon, it's more efficient to first analyze the one with the fewest variables – to observe any cause and effect.

Misquoting Occam's razor applies it incorrectly, which can be misleading when little research is done otherwise. Scientists know there is more to proving a theory, or explaining a curious phenomenon, than just accepting the simplest explanation. It isn't enough to merely latch onto Occam's razor and use it wrongly and out of context. That can create false and useless conclusions.

It seems people use the Occam's razor argument in a way that says the simplest explanation is the one that fits most easily and most comfortably within the box of their belief or understanding about reality. Anything outside that box is not as simple and easy for them to grasp. Therefore, it is not the right explanation.

The result is that laypeople can make poor decisions based on misinformation and falsehoods. That causes mistaken beliefs and lifestyle decisions which carry a lot of weight in society. This, in turn, causes weak government and a decline of public well-being.

Carl Sagan's Baloney-Free Diet

In Carl Sagan's book, *The Demon-Haunted World: Science as a Candle in the Dark,* Sagan offers nine tools to protect scientists and others against becoming victims of falsehoods. He views Occam's razor as a rule-of-thumb that advises choosing the simpler of two hypotheses that offer equally good explanations when comparing available data.

That means starting with the simplest hypothesis to analyze for correctness – not to immediately accept the simplest as already correct. For Sagan, Occam's razor does not stand alone as the only tool for deciding the validity of a hypothesis. It's just one of nine tools he suggests. Sagan offered great practical wisdom in these tools. He included other tools as well to help determine if even the simplest hypothesis is a correct explanation.

Occam's razor is not a gauge of correctness by itself. There's still a need for repeatable experimentation, logic, comparisons with different hypotheses, etc. It doesn't matter if you're trying to explain a UFO sighting, natural phenomenon, novel health treatment, or a clean energy breakthrough. You must do the necessary amount of work before accepting an explanation as correct. Even artificial intelligence can produce false information.

Finding Truth from Gut Feelings

A recent and growing amount of research suggests there is a connection between microbes in the intestinal gut, and higher brain functions such as intuition. That reinforces what people have said for centuries about gut feelings. If you don't feel right in your heart or gut about something that is supposedly true, maybe it's false. When analysis and gut feelings line up, it's easier to accept something as true. When they don't, it can be worth a little more research to find out why.

The main point about Occam's razor and our gut feeling is that we all make many choices in life. Correct choices require correct information. How often have you received information that did not feel right and caused you to doubt it? That's happened to me many times.

Upon further research, I discovered that authorities or experts who presented the information were incorrect. After I challenged their statements, based on my careful research, they retracted their original statement. That wouldn't have happened without me questioning their information!

Modern Philosopher's Legacy of Truth

Bertrand Russell was a British historian, mathematician, and philosopher. He was passionate about truth, love, and justice. He won a Nobel Prize in Literature. BBC's program, *Face to Face*, asked Russell what lessons he would like to share with future generations. His reply was both practical and compassionate.

The intellectual part, for him, encouraged people to focus only on facts and truth when studying anything. He urged that beliefs and considerations of social benefits should not become a diversion. His moral lesson was that wisdom comes from love and it is foolish to indulge in hatred. To Russell, charity and tolerance were essential for human life to continue on Earth.

Why go to extra effort to find out what is true? Have you heard the old saying, "Believe nothing of what you hear and only half of what you see"? By "old," I mean that similar statements were around long before Edgar Allan Poe, who often gets credited for that statement. The search for truth goes back to ancient Greek philosophers and even earlier. In essence:

> **Truth is essential for relationships and connection at all levels. Reintroducing and reclaiming the veracity of truth is vital for the next step of civilization to happen. Without truth, there is no trust. Without trust, there can be no love.**

Maybe you're merely taking a true-or-false test. Perhaps you are investing money and want to know the real risks and potential profit. Possibly, you are not feeling well and want to know the truth about your health from a qualified health-care practitioner. You might be selecting a school for your children and want to know about bullying there. The more important the decision, the more valuable truth becomes and benefits you.

Do you agree that knowing truth is about enjoying your "life, liberty and the pursuit of happiness," as America's founders put it? Knowing truth gives us power—the power to enjoy better health, more happiness, love, and prosperity, for ourselves and loved ones. When someone doesn't tell you the truth, do you think that person might be trying to have power over you? It might be for political or financial reasons, or to gain something else from you. Maybe they believe you can't handle the truth!

Mainstream News Needs a Higher Truth Standard

Have you noticed how influential news media struggle with reporting even the simplest facts? Major news outlets need to maintain a higher standard because the public counts on them to share the truth. Mistakes in reporting news have become rather absurd. Worse yet is that coordinated lying, or spinning the facts in news reports seems almost routine. This is actually old news!

People tend to believe major news outlets, but that is the very least of the problem. Before continuing, a couple of definitions are in order:

- A *fact* is true because it is real, or has happened.
- An *opinion* is an assessment, or something that people believe, regardless of whether it's based on fact.
- The *news* is a report of one or more events in society, the world, political campaigns, sports competitions, the weather, and so on.
- *Journalism* is more about reports on the interaction of events and people, or ideas and other facts, to provide useful insight and information to society.

Reporting an incorrect game score is a forgivable human error. It's not a human error when news outlets deliberately misinform their audiences to achieve a hidden motive or agenda. Granted, the definition of news doesn't include responsibilities to check the facts of everything printed or broadcast as news. News outlets might merely be reporting what they consider newsworthy from a person that is well known. The person's statement might contradict easily verifiable facts, yet news outlets report it anyway, and viewers are still affected in a negative way.

Journalists have a great responsibility to report and analyze news accurately. At least it's in their best interest. We like to think journalists are a reliable source of information. What's an example of when you believed news was factual, yet found out later it was merely an opinion *not* based on facts? Is there anyone who has not had that experience? It can be challenging to know if news is true.

Do you ever take shortcuts? Do you rely on others to tell you what is true? Why do we do that? Maybe we think it can save time if we view someone as a trusted source of information. Maybe the truth we are trying to understand is just too complicated.

Do you have the time, interest and knowledge to determine what the weather will be tomorrow? I certainly don't, so I look up weather predictions on the Internet. However, many bigger issues don't have such reliable and easy answers. These issues affect us in potentially life-changing ways.

Every day, people say things that contradict valid evidence, and they refuse to accept the facts. They are overly consumed by a passionate belief in what they wish was true. No one benefits from that kind of simplistic acceptance of wrong information.

Opinions, rumors, and beliefs *can* be right, but they need to be verified to have real power. They can motivate people and create momentum by giving people an empowered understanding using evidence-based, common standards. Real power comes only from the absolute reference point of truth. The clearer and more exact our understanding of truth is, the more powerful we will become.

Science of Lying

- How did we get to where so many people value truth less than before?
- Have you heard someone say something, and then almost immediately deny saying it?
- Have you accepted something as true, merely because many people have said it for a long time?
- Does truth still matter to you?

Research at UC Berkeley revealed that the prefrontal cortex works harder when a person is lying. [See "Liar, Liar, Brain on Fire" by Janice Gee, Fall 2005 www.ocf.berkeley.edu.] According to their research, the brain has to work harder to create a story that is not true. Other research, at UCLA, indicates that lying results from a lack of connection between the prefrontal cortex and brain areas that are involved in altruism and empathy.

Another scientific study about lying showed that a person's brain can desensitize to lies that they tell, even if their lies get bigger over time. The news media widely reported results of that scientific study published in *Nature Neuroscience* on October 24, 2016. Joseph Goebbels, Adolf Hitler's chief of propaganda, knew of this fact. He said, "If you tell a lie big enough and keep repeating it, people will eventually come to believe it."

Does that sound familiar? No one is immune from this tactic coming from politicians, corporations, or other people we think we can trust. It's not just the lying, but also the quick and easy acceptance of lies that has become so damaging to society.

The Art of Misdirection

Do you like watching magicians? One technique of magicians is misdirection. A magician will attract your attention to their right hand, while they do something sneaky with their left hand. We see this frequently in misleading articles and other ways. The "Bait and switch" tactic is similar and common.

Large fines for major telecommunications companies and other corporations have resulted from them billing customers for things they did not order. It takes extra alertness to catch these things. When a magician entertains you, it's fun. When someone lies to take advantage of you, it can be devastating in many ways.

Diagnosing the Fake News Epidemic

Fake news is a growing and pervasive problem for every area of life. Does part of this come from our historic love of *tall tales* and other fictional entertainment? What's your favorite tall tale? One of mine is, "I caught a fish so big that it took four strong men to pick up a photograph of it to load onto a truck." We're not talking about that kind of story, although a lot of news seems just as exaggerated. Many people want it to be true, so they believe it despite any valid evidence to the contrary. Tragically, though, fake news is contributing to the downfall of human civilization.

Misinformation gets spread by those who wish to gain profit, power or some other advantage over you. Lying, fake news, hearsay, rumors, and misinformation, have contaminated every form of media and are part of the entire history of humanity. Rampant propaganda in wars, politics, and businesses has been around for as long as those activities have existed.

Tabloid newspapers, mainstream news, and even social media influencers have made fortunes spreading fake news. According to Robert G. Parkinson, an assistant professor at Binghamton University, even Benjamin Franklin, one of America's most respected founders, used fake news. Parkinson authored *The Common Cause: Creating Race and Nation in the American Revolution.*

As one example, Parkinson shared that Franklin once falsely accused Native Americans of atrocities and bragged about it. Big corporations routinely use fake news, falsified research and paid third-party experts to win customers and increase profits. They make false claims about product safety and exaggerate benefits to please customers, investors, and government regulators as routine business strategies.

That's why it became necessary to make laws about "truth in lending," "truth in advertising," "truth in labeling," etc., where truth is deemed critical to protect consumers. Major corporations still break these laws because the fines and slap on the wrist are small risks compared to huge profits they can make through their dishonesty and blatant criminal activity. Protecting ourselves from fraud and other atrocities is one of the responsibilities that "We the People" delegated to government, but our government commits fraud against us as well.

Why are Misinformation and Fake News Bad?

People have begun to view truth, beliefs, and opinions as equal to each other. They can be, but only when they are all backed up by verifiable facts. It seems, though, that people are starting to doubt that there is any truth at all. Global chaos and incoherence have lowered the understanding and value of what truth really represents. It's a vibratory disconnect which shatters all human beings at a fundamental level.

A singer can shatter glass with their voice. However, flaws in glass are what cause it to break, not the vibrations themselves. Misinformation creates extreme incoherence that can shatter the windows of our perception of truth. Flaws in our social systems, and us, cause this. Results of misinformation can be as big and tragic as genocide and global destruction, or as small as decisions in our daily lives, which can also have big consequences.

We must understand the destructive nature of misinformation infecting the world. Truth is necessary to rebuild a healthy system, and must be valued. All systems break down in society when we stop valuing truth. Lies and fake news are destroying us. If you spread information that our planet is not at risk, you will kill it by believing in a falsehood. Only truth can generate the necessary passion for saving humanity.

Lying as a Weapon of Mass Destruction

A covert misinformation campaign is underway. It's designed to manipulate and control individuals and the mass of humanity. It has become a weapon of mass destruction throughout our world! Exposing masses of people to misinformation happens in nanoseconds now, and it's not random. Fake news is not coming only from entrepreneurs creating their income and pranksters playing a joke on gullible people. How does fake news spread?

- Public relations (PR) companies and governments have mostly been the source of false news until recently. They have staged deceptive visual imagery, and planted fake news stories in mainstream and social media to manipulate the opinion of the public and other stakeholders.

- Politicians spread misinformation that will get them votes or campaign donations. Their supporters believe it no matter what the truth and facts are. The U.S. Supreme Court ruled that it is legal for politicians to lie, all in the name of "free speech."
- Fake content is created on social media for more profit.
- Fake news bombards the public, and people find it easier to accept a lie rather than verify the truth.
- Hackers can infiltrate news sites and falsify information. However, hacking is illegal, whereas fake news is not.

The Cancer of Misinformation

Does it seem extreme to call misinformation a cancer? Well, cancer in the body *is* misinformation. Even human cells have an innate sense of truth. Some cells forget their nature and become cancerous. They become a deadly source of misinformation and try to integrate into our bodily systems. A body resists and rejects these cancerous cells until they multiply enough to overwhelm healthy cells and destroy our body.

If something does not abruptly stop the spread of this kind of bodily misinformation, it takes over and destroys the integrity of our biological system. Misinformation ruins the cooperation, resources, expansion, and growth of a healthy system. Fake news and untruths in any form are therefore cancers in society.

Knowing how to investigate truth is required for you to know truth. Systems and sources of truth and facts are breaking down. You have to want truth and not merely accept that truth can exist. Having an intention and commitment to know truth are crucial first steps. You are designed to know truth. At the level of your DNA, and beyond the deepest quantum level of your being, you resonate with truth when you encounter it.

A Cure for Misinformation

Does it seem like we need a holistic way to let truth triumph? National Public Radio (NPR) offered *A Finder's Guide to Facts* in a December 2016 online article. These practical guiding tips require varying degrees of time and effort.

If truth is important to you, then NPR's guide is valuable. Every culture on Earth has had brave men and women of higher conscience who most likely talked about the importance of truth. The Greek philosopher Socrates suggested a "Triple Filter Test." He used it to determine if news from someone else about Socrates' friend was worthwhile.

Socrates asked the person if the information was true, if it was something good, and if it was going to be useful to Socrates. He advised asking these questions of anyone sharing information with someone about one of their friends. Socrates decided that the news was not worth his attention if any of the answers to his questions was "no."

"DO YOU WANT TO KNOW THE TRUTH?" That is the big question. If truth does not matter to you, then others will deceive, rob, cheat, exploit, and control you.

The moment you start respecting truth, simultaneously LOVE and coherence begin to develop. This creates even more incentive to be passionate about truth. The value of truth must be elevated so that people are inspired to pursue truth. That's why we need to give truth a chance. It creates another upward spiral of mutually reinforcing influences. A few other solutions to the pervasive problem of declining truth are in later chapters.

It is good advice to not take anything for granted or believe anything at face value. It doesn't matter if it's a statement by a politician, a bill from a big or small business, election results, or even sports scores. Think twice, especially if you intend to make an important choice based on the information you receive.

Before you offer something as true, do a little research. That does not mean ending the practice of speaking truth in a kind way.

It does not mean you stop protecting someone's feelings with what might be an overly optimistic statement. It is more about discovering truth and facts related to your opinions and beliefs. What you believe is the basis for decisions you make.

Is Truth Compatible with Freedom of Speech?

How can anyone make informed choices based on lies and other misinformation? A new law could be created by citizens to require politicians, news media, businesses, and PR companies to inform people when their statement is false. They could still lie, but would be legally required to indicate it is false.

It's similar to other labeling laws. For example, you can sell shirts made of many kinds of material. However, if a shirt is 100% polyester, you cannot legally label it as 100% cotton, at least not in the U.S. A person who buys that shirt might believe it feels like cotton or doesn't care, but they can make an informed choice.

Costly fines for violating this law would be a big incentive for compliance. A "Truth in Speech" labeling law would reasonably protect both constitutional rights of free speech, and the rights of voters and consumers to have facts. It even protects those who choose to believe something that is untrue.

This kind of labeling also gives voters a chance to protect themselves from the unhealthy effects of a lie. For example, let's say a politician lies to voters to evoke feelings of hatred toward a rival candidate. If they are successful, then many voters will be forced to endure the self-damaging effects of feeling hate.

Negative emotions can have stressful and damaging effects on the human body. They create health risks that could shorten one's life and even be deadly. If the voters know the statement is a lie, then they have more choice and the ability to protect themselves from harm. Imagine the benefits that greater truth and positive emotions would have on lowering health care costs!

Consumers often choose to ingest poisons of various kinds, such as tobacco and alcohol. At least they know what is in the product, and the risks. They can choose to consume them or not. Voters who ingest toxic information from someone who lies to them cannot make an informed choice.

Do we need a label for politicians to wear that says, "Warning: Beware of Toxic Lies"? In chemistry, lye is highly toxic and causes chemical burns. It can be fatal if consumed. Have you ever been "burned" by someone who told you a lie?

Can you imagine how politicians, corporations, PR firms, news media, and their respective lawyers might resist such a law? Would you find it entertaining to hear these people argue that, "If we have to admit that a statement is a lie, then that defeats the purpose of lying in the first place"? Would you expect that a combination of fines, embarrassment, and the sheer nuisance of complying with such a law would be helpful?

Speaking the truth is so much more cost-effective. Truth is more powerful when shared than kept to ourselves. It is common to all of us. If "We the People" really have power, then would you agree that it comes with great responsibility? Knowing and sharing truth is a vital and urgent responsibility for the entire world.

Truth is getting even more critical as artificial intelligence (AI) replaces human beings to achieve an increasing number of tasks. AI scans vast data resources to provide useful output. It even throws in some of its own randomness to seem more "human". That, combined with false data, misinformation, propaganda, etc., is a perfect storm for creating false, useless, and even dangerous output from AI. That scenario is called, "AI hallucinations"! That's not a tour joke. It's true. Now, would you want to get a medical procedure or ride on a commercial flight, that's controlled by someone who is hallucinating? Probably not!

Seizing the Initiative

As I journeyed down my paths to develop my full potential and learn about solutions to world problems, some things made a big impact on me. Along the way, I had to learn what was true or false. I also looked for new and better ways to present truth to the masses. Some of the most exciting and rewarding experiences of my life resulted from "seizing the initiative." That includes using the power of the people in a form of self-governance called ballot initiatives. The next chapter shares what "Power of the People" can achieve. Are you ready to explore such empowerment?

Summary

Knowing and sharing truth is vital for our survival. Lies, false information, and propaganda are damaging to all levels of life and our civilization. That is even more dangerous as artificial intelligence increasingly guides, augments, and even takes over human activities. Lack of truth and accurate data can cause AI programs to produce AI hallucinations!

Lies and misinformation are both a cancer and a weapon of mass destruction. The incorrect use of Occam's razor contributes to that damage. Beliefs based on truth can be powerful stepping-stones for you to take initiative.

By discovering truth outside and inside you, that bright spark of inspired initiative can further ignite your highest and greatest purpose and successes in life. It requires making an effort to find truth and be discerning along the way. Truth is your innermost nature. Use your gut feelings too, when faced with important new information, especially if you intend to make decisions based on the information you acquire.

Your decisions determine what kind of life you have. My decision to "seize the initiative" took me on some thrilling journeys. We'll see what happens when important and unfamiliar truths are shared with politicians and the People.

3

Light Up the People

"...the people are the only legitimate fountain of power..."

– James Madison

As your tour guide, I will now review with you some history about power and political initiatives. Power is neither good nor bad. It's merely your ability or competence to do something. You can increase power through leverage to be even more empowered. Buckminster Fuller urged people to see what needs doing, and take initiative to do it, even if no one else does. By doing that, he believed the "real you" would be brought out.

How can you use your power to achieve anything? Initiative is the key! Fuller took the initiative to design innovative solutions for global problems. Legendary inventor, Nikola Tesla, took the initiative to create technologies that advanced societies worldwide—possibly more than any other human being.

Over 100 years ago, Tesla invented a new alternating current motor that powered up our world with electricity. He gave us far greater leverage to advance in every way. Like you and me, he also made mistakes. In his autobiography, *My Inventions,* Tesla told stories of his youth. He wrote about how he was better at taking apart his grandfather's clocks than he was at putting them back together again.

Have you ever taken something apart and then not been able to put it back together? I've done that often. It's a common trait of inquisitive youth. Adults who see a need for change do that too. It doesn't mean everyone who does that achieves breakthroughs or worldwide fame. Which one or more of the following scenarios applies to you?

- You take it apart and don't know how to put it together.
- You take it apart and *know* how to put it together.
- You take it apart, put it together, and make it better.
- Like Tesla, Steve Jobs and other innovators, you put new things together that were thought to be impossible, and you make our world better!

America's founders took apart an oppressive government and designed something better, even though it still had weaknesses. Our natural curiosities are often seeds for our life path and our successes. They can happen at any age. Do you remember any transforming experiences from your youth?

Tesla's curiosity empowered him to invent not only a vastly superior alternating current motor, but also radio, robotics, wireless energy transmission, and other world-changing technologies. His inventions increased power of others. Even in the early 1900s, Tesla described what we now know as cell phones, Internet, GPS and other technologies that would have seemed impossible.

What makes anyone capable of inventing something out of nothing? Leonardo da Vinci, the great Renaissance inventor said:

"Learn how to see. Realize that everything connects to everything else."

Another value of knowing truth is that it enables us to understand what we are seeing and know that it is real. On that basis, we gain a firm ground for our intentions and initiatives. Steve Jobs claimed that once you discover and learn the truth of your power, you will never be the same. He was talking about a change that will bring you more in line with your true self. Jobs, Tesla, and Fuller were legendary inventors and innovators.

Everyone has similar innate abilities. These include knowing truth, taking initiative to achieve something, and committing to actions toward that achievement. All of those abilities have power. Combined with leverage offered by technologies, systems, and tools, knowledge of truth gives you enormous power.

- Recognizing your power turns it on.
- Taking the initiative gets your power engaged.
- Using your power for good gives it a higher purpose.

America's founders recognized the power of noble intentions. **Great power comes *only* through noble intention!** That's a bold declaration for the new civilization we are creating. It's not a mild suggestion of something to consider. America's founders had to be that bold in their day. We need to be even bolder in our day. They chose to create a new society and government. Even at great personal risk, they committed to that path. They made history for themselves and a future for us. We must boldly choose to realize the fullest potential of the government and civilization that they envisioned. It's vital for our survival.

Everything achieved throughout human history started with these basic, but powerful, steps of learning truth and committing to actions with persistence. Every country and community had people who inspired or benefited others with their initiatives. These have been scientists, inventors, prophets, activists, athletes, revolutionaries, and more. They chose to move forward in ways no one else had before. A simple idea or goal sparked a fire in one person and it spread to advance and uplift human life.

Engaging the Power of the People

How do *you* share truth with others? I love the smell of ballot initiatives in the morning. The sweet fragrance of self-governance is nourishing to me. An "idea whose time has come" is energizing.

Engaging the power of the people is an extraordinary way to offer solutions to societal problems. My experiences with several ballot initiatives were from 2000-2010.

Some of my comments in the next two chapters might seem harshly critical and disrespectful. However, that was before I learned deeper reasons for the behavior of politicians. Since then, I have reached conclusions that will enable you to gain a deeper understanding of why politicians are inherently unable to prevent or solve problems. By knowing that, your success and peace of mind in implementing your solutions will increase.

What is the Ballot Initiative Process?

The ballot initiative was a game-changing innovation in self-governance derived from the right to "...petition the government for a redress of grievances," as stated in the First Amendment to the U.S. Constitution. That means you have a legal right to bring a grievance to your government and demand its help to solve a problem, without fear of reprisal from the government.

America's founders might not have foreseen how important those eight words in the Bill of Rights would become. However, they predicted inevitable corruption of government. Legally binding ballot initiatives also use powers implied by the Tenth Amendment, which states, "The powers not delegated to the United States by the Constitution, nor prohibited by it to the States, are reserved to the States respectively, or to the people." Ballot initiatives are "practical politics" in action.

"To the people..." Do you get it? These visionary statements were a gift of great power to us from America's founders. They also reflect power that we have, even outside of government. It helps citizens achieve crucial public policies, without interference from politicians, through *direct democracy.*

At least 75% of the U.S. population lives in twenty-four states and the thousands of counties, cities and towns where voters have this right to propose and enact laws. Ballot initiatives empower citizens to propose a law, usually on a single subject. It requires collecting signatures of voters to put a measure on the ballot in an election. This process of self-governance allows people to make laws with their vote. Government officials usually cannot veto a ballot measure law that is enacted by the people.

Thomas Jefferson, James Madison, and even Plato knew that politicians elected in a republic were likely to get addicted to power and act more in their own interest than that of the public. The populist movement of the late 1800s saw that America had reached the undesirable political condition predicted by Jefferson and Madison. To achieve their crucial reforms, populists enacted the initiative process. They created a tool to help fix what is wrong when the government would not. They empowered themselves with the tool of direct democracy to better serve the People.

"Plan B" from America's Founders

The ballot initiative process is a Plan B for democracy. All of the major problems we face, as individuals and as a society, are the result of our individual and collective lack of vigilance in taking responsibility for our nation, our planet, and ourselves. Our wise founders knew that Americans' lives shouldn't be the sole responsibility of an elite group of elected representatives. James Madison acknowledged this by declaring that, "...the people are the only legitimate fountain of power..." [*Federalist Paper No. 49*]

We the People have the power to create the lives we want. We gave elected representatives limited powers in various areas. When they don't do their job well, we have to think how we, as the "original authority" and "fountain of power", can improve government. "The People" can make laws too!

Citizen-sponsored ballot initiatives have played a vital role in progressive American public policy for over 100 years. They have achieved many of the most crucial reforms in U.S. history. Their successes include women's right to vote, tax-reduction, child labor laws, and environmental protection laws.

News media give a lot of attention to citizens who become lawmakers with ballot initiatives. Special interests don't control these citizens as much as they do our representatives. A formal ballot initiative can achieve many goals. It puts politicians on notice. It tells them that people are not satisfied with the unwillingness of government officials to consider innovative, yet proven solutions. That includes issues as large as climate change, global air pollution, pandemics, mass gun killings, and other related problems. Citizen lawmakers vote on these new laws, so they are more likely to study and discuss the ballot initiative issues.

Increasing Truth through Public Policy

Could a ballot initiative increase the honesty of politicians? Over 25 years ago, I drafted two ballot initiative concepts to deal with fake news and lies by politicians. They would be like truth in advertising and labeling laws to enable better informed voters.

You could consider one of the proposed ordinances as sort of a "Truth in Speech" law. It aimed to reduce lying by government officials. The draft ordinance declared that (in summary):

- All powers of every branch and level of government are derived from the People.
- The court of public opinion is, therefore, the highest court in the nation.
- Government officials are in service to the People.
- The People deserve truth from their public servants to make the best choices for society and themselves.

The binding part of this law would have required that anyone addressing the public on behalf of government would be under oath to tell the truth, and therefore subject to punishment or fines for perjury if found guilty. The other proposed initiative would have required labeling, or a "Truth in News Reporting" law, so that consumers and voters would know if public relations companies or news corporations had falsified news on media outlets.

Jumping into the Deep End of Direct Democracy

My involvement with ballot initiatives began with a wake-up call. I was a U.S. Senate candidate with the Natural Law Party in 1998. It was shocking to see how third-party candidates were treated unfairly, and cost-effective solutions were ignored. One of my platform points was to provide ways to reduce stress among high school students as a means to prevent violence. Five months later, the deadly Columbine High School shootings happened.

Five months after that, President Clinton held a summit on school violence. The main conclusion by attendees was that stress was the number one cause of violence in schools. Students know this too. And yet, the Clinton-appointed summit moderator steered the focus back to the topic of gun control. Consequently, only superficial approaches to preventing violence are still used. Tragic killings continue in schools and elsewhere.

During my campaign, I saw a two-page article in the Denver Post about a ballot initiative that proposed regulations for hog farming in Eastern Colorado. That was my calling – not hog farms, but ballot initiatives. Soon after, I realized how ballot initiatives could introduce cost-effective, preventive societal solutions. The Institute of Initiative and Referendum, and the book titled, *Citizen Lawmakers,* by David Schmidt, were my best resources. The book was thrilling in describing the way ordinary citizens bravely took on city hall or even state government.

My first attempts at direct democracy were in Colorado. At the Colorado State Capitol, I answered questions from legislative staff for several hours about my three initiatives. The first addressed state budgeting. It introduced the idea of "giving priority in all state budgeting to proposed programs and solutions that best satisfy the criteria of being efficacious and cost-effective" in achieving the mission for respective departments and agencies requesting budget funding. That was 25 years before Elon Musk's reckless and self-serving Department of Government Efficiency (DOGE) assaulted the Federal government and did untold damage.

My second initiative would have required health insurance carriers to cover all health care services, including stress reduction methods, which were efficacious and cost-effective according to peer-reviewed research. The elected politicians ignored these commonsense policies. I explain why in Chapter 6.

Truth in Labeling GMOs

I authored a third statewide ballot initiative to require labeling of genetically modified food, aka GMOs. Food safety champions such as Laura Ticciati, Dr. John Fagan, and lawyer Steven J. Druker, inspired my interest in the GMO controversy. Ms. Ticciati, author of *Genetically Engineered Foods: Are they Safe? You Decide,* and her organization, Mothers for Natural Law, worked tirelessly to raise awareness about the issue. Molecular biologist, Dr. Fagan, took a stand against genetically engineered food. In an October 11, 2000 article, *The New York Times* reported about Dr. Fagan that,

> "...In 1994 he turned down a $614,000 research grant from the N.I.H. to study gene therapy because of concerns about the 'dangerous consequences' of manipulating human genes. And two years later, he founded a company that tests crops for genetic alterations, giving processed-food makers the option of steering clear of biotechnology ingredients."

Attorney Steven Druker, along with a "Who's Who" of public interest organizations, scientists, and religious leaders, filed a lawsuit against the U.S. Food and Drug Administration (FDA) in the late 1990s. It forced the FDA to release memos showing the agency ignored its own scientists' warnings about genetically engineered foods. Druker authored the 2015 book, *Altered Genes, Twisted Truth: How the Venture to Genetically Engineer Our Food Has Subverted Science, Corrupted Government, and Systematically Deceived the Public*. The book's foreword was by the renowned primatologist Dame Jane Goodall. She praised it as "without doubt one of the most important books of the last 50 years."

Blazing a Healthier Food Trail in Trail Blazer Land

Concerns by a spectrum of consumer rights stakeholders called for a new approach to raising awareness about the issue. It was ripe for a ballot initiative. I shared my draft statute on GMO labeling with Donna Harris, an Oregon activist. She was a brave, impassioned, and concerned single mother. She, and a legion of committed volunteers, collected petition signatures to qualify the initiative for the Oregon statewide ballot in 2002 as *Measure 27.*

Donna was a champion. Her determination inspired dozens of major organizations and businesses. She and other supporters faced a daunting adversary. Monsanto, the notorious global polluter, led an army of well-funded entities and unleashed a firestorm of misleading propaganda. The campaign started a few months after Monsanto lost a lawsuit in Alabama. *The Washington Post* reported on February 23, 2002:

> "An Alabama jury yesterday found that Monsanto Co. engaged in 'outrageous' behavior by releasing tons of PCBs into the city of Anniston and covering up its actions for decades, handing 3,500 local residents a huge victory in a landmark environmental lawsuit." It further stated that "the jury held Monsanto and its corporate successors liable on all six counts it considered: Negligence, wantonness, suppression of the truth, nuisance, trespass, and outrage. Under Alabama law, the rare claim of outrage typically requires conduct '...so outrageous in character and extreme in degree as to go beyond all possible bounds of decency so as to be regarded as atrocious and utterly intolerable in civilized society'."

Goliath Shakes in His Boots

Measure 27 was an epic David vs. Goliath battle over the right to know if any food contained potentially hazardous genetic modifications. It became an astonishing success for opponents of GMOs and supporters of labeling. It drew worldwide news media attention to the issue like never before. It rallied a broad coalition of stakeholders who were concerned about the risks of unlabeled and untested experimental food.

The focus on a single state was a breakthrough. It set the stage for similar initiatives to follow. Mel Bankoff, a visionary organic food entrepreneur, and other large players in the natural foods industry supported the "Vote Yes on 27 Campaign". My role was as a consultant for the campaign and to help obtain and coordinate endorsements. Many of those who endorsed *Measure 27* applied their unique and creative strategies to promote it. United Natural Foods, Inc. placed full-length banners on their company semi-trucks to promote *Measure 27*. The groups included:

> Consumers Union, Center for Food Safety, Democratic Party of Oregon, Friends of the Earth, Greenpeace U.S.A., Oregon League of Conservation Voters, Oregon Physicians for Social Responsibility, Oregon Rural Action, Oregon Tilth, Organic Consumers Association, OSPIRG, Pacific Green Party, Sierra Club (Oregon Chapter), Union of Concerned Scientists and United Natural Foods, Inc.

Paul McCartney's Endorsement – "Let it Be" Labeled

During the campaign, I heard that Paul McCartney would be doing a concert in Portland. I sent a message in five directions to request his endorsement of *Measure 27* during his visit. The concert came and went without any response from McCartney. It was a long shot and seemed to be unsuccessful.

One week later, Mel Bankoff called me and excitedly said, "Guess who I just talked to? Paul McCartney!" Mel described that McCartney had heard about our campaign and wanted to support it. His late wife Linda had been a dedicated advocate of labeling GMOs, and of healthy food in general. Paul and Mel recorded a 30-second audio that McCartney ended with, "so, *let it be* labeled," – referring to the Beatles' song. Mel and others then bought airtime on radio stations to play McCartney's endorsement.

Measure 27 did not pass, however. Food industry opponents of the initiative spent almost $7 million to defeat it, and outspent our campaign 67-1. That was my first exposure to how a major industry can buy an election outcome with a well-funded flood of lies and misleading propaganda.

This first GMO labeling ballot initiative made headline news worldwide. Agence France-Presse summed up the defeat of *Measure 27* with, "The global food industry is breathing a sigh of relief," or something similar. The Associated Press, as I recall, stated that *Measure 27* was "the initiative of most interest" during that election season. So, it was still an unprecedented success. Soon after, "Non-GMO" labels began appearing on foods everywhere!

Safety through Peace – Initiative 101 and Counterterrorism

Jumping in the deep end of challenges has always been my nature. So, I went big! My new proposed ordinance drew on some pioneering research about reducing stress-related social indicators such as crime, violence, and even terrorism and war—globally! I called it *Safety through Peace.*

I filed my draft ordinance with the City and County of Denver in early December, 2002. On December 10, Denver City Attorney David Broadwell, and city council Director of Staff John Bennett, reviewed it with me in a mandatory televised public meeting.

The timing seemed perfect. Momentum was increasing for the U.S. to wage war in Iraq. On February 4, 2003, the Denver City Council passed a controversial non-binding "Resolution No. 10, Series 2003 – Opposing War with Iraq without Full Congressional Debate and Authorization." Impassioned and lofty speeches urged passage of the resolution. One of its sections encouraged me. It said, "Now, Therefore, Be It Resolved by The Council of the City and County of Denver:

> Section 4. That we urge the citizens of Denver to contact their federal representatives to urge peaceful solutions to international conflicts."

Two months before, I had proposed "peaceful solutions to international conflicts" as well as to community violence and crime. The atmosphere was ripe for my proposed ordinance, or *was* it? My proposed law, if passed, would have amended the Code of Ordinances of the City and County of Denver with an ordinance titled, "Increasing Public Safety by Creating Peace". It explained how stress builds up in society and causes a variety of societal problems, including terrorism. Creating coherence in the collective consciousness and reducing societal stress were the keys. The following question was put to voters in November 2003.

> *"Shall the voters for the City and County of Denver adopt an Initiated Ordinance to require the city to help ensure public safety by increasing peacefulness—that is, by defusing political, religious and ethnic tensions, both locally and globally—through the identification and implementation of any systematic, stress-reducing techniques or programs, whether mental, physical, etc., that are (1) scientifically shown to reduce society-wide stress, as measured by reduced crime, accidents, warfare and terrorism, and also (2) shown to be of net financial benefit for the city?" Yes__ No__*

Support from Leading Experts

An impressive list of experts had already supported this kind of approach of reducing society-wide stress as a proven strategy for reducing crime, violence, and terrorism. These included:

- David Edwards, Ph.D., Professor of Government at the University of Texas at Austin and one of America's foremost political scientists. Millions of college students used his textbooks;
- Peter Salk, M.D., son of Dr. Jonas Salk and vice president of the Jonas Salk Foundation, who was leading an international research effort to develop an AIDS vaccine; and
- John Davies, Ph.D., a leading researcher at the University of Maryland's Center for International Development and Conflict Management.

Other leaders had also supported this approach. These included policy experts such as Professor Ved Nanda of the University of Denver, and Dr. Martha Urioste, a pioneering educator, Colorado Women's Hall of Fame inductee, and long-time advocate of world peace.

Another supporter was Major General Kulwant Singh, a highly decorated 30-year veteran of counterterrorism in the Indian Army. He had arrived at the National Press Club in Washington, D.C. on the morning of Sept. 11, 2001 to introduce this proven approach to counterterrorism. His presentation was announced days before but it didn't occur due to the terrorist events of 9/11.

Major General Singh enthusiastically supported my *Safety through Peace* initiative. It could have implemented the solution on which he was to speak. On May 15, 2002, he informed me via e-mail that he had read my ordinance and was "passing on the entire correspondence to all Regional Directors (Generals and Brigadiers) in each Indian state as well as to the Deputy Defence Ministers in each continent for wide circulation." Naturally, I was both thrilled and honored by his support.

Colonel Brian Rees, Medical Corps, US Army Reserve, was also a strong supporter of Initiative 101. He was the author of the book, *Terrorism, Retaliation and Victory – Awaken the Soul of America to Defeat Terrorism without Casualties.* An Army recruiting video featured Colonel Rees as a top Army medical doctor.

Endorsement from Holocaust Museum Benefactor

Another valuable endorsement for Initiative 101 came from Jeffrey Abramson, a philanthropist and a partner in The Tower Companies. His award-winning, family-owned real estate development company in Rockville, Maryland is a green building leader. The Abramson Family Foundation philanthropy includes generous sponsorship of the arts, science, education and medical technology, religion and the humanities, as well as world peace programs.

They guaranteed the loan for construction of the Holocaust Museum in Washington, D.C. They also supported a peace project in Washington, D.C., which reduced violent crime by over 18%, and total crime by over 23%. That was in 1993, at the same time as Denver's infamous "summer of violence."

Obviously, Jeffrey was a busy man. Yet, he took time to write a personal letter to the Denver City Council, strongly supporting *Safety through Peace* - Initiative 101, and urging *their* support. Mr. Abramson knew of Denver's 50% increase in violent deaths that year, and of Denver's dire financial straits. He confidently assured the councilmembers that the *Safety through Peace* ordinance would bring great rewards.

You'd think that city leaders would welcome input from such outstanding experts and humanitarians on critical issues such as crime, violence, and terrorism. One bright light was Denver City Councilwoman Happy Haynes. She filed bill CB 587-03 for the ballot initiative to ensure a first reading by the city council. Shockingly, Haynes was the only member who voted for it.

Surprising Hostility from Politicians

Both old and new city council members asked at least some legitimate rhetorical questions. Other members ranted, whined, or made hypocritical comments. They appeared to be merely saying whatever they thought would appeal to their base of supporters, while ignoring extensive research supporting the premise of my proposed ordinance. At least one council member claimed it was too easy for citizen-sponsored measures to get on Denver's ballot.

Ironically, at that time, it took almost 2,500 signatures for any citizen to get a single ordinance on the ballot. By comparison, only 100 signatures were required for any of the eleven district city council members to get *themselves* on the ballot. Once elected, they could pass as many ordinances as they could get through their legislative process during their terms in office. It's unlikely that any of them ever did the hard work of direct democracy.

City council members typically got on their soapbox to decry horrible violent crimes. It didn't matter whether it was about the Denver police killing Paul Childs, a developmentally disabled 15-year-old boy, the worst summer violence in ten years for Denver, or school shootings with assault rifles. Gun rights vs. gun control remains a never-ending argument. However, politicians ignore that, according to the FBI, 450 U.S. children are killed each year by their own parents without using assault rifles or any guns at all.

Getting Peace on the Ballot

After turning in our petition signatures, the Denver election office reviewed their validity. They informed me that we were thirty-one valid signatures short. Rules for ballot initiatives vary greatly between city, county, and state levels, in different places throughout the U.S. In Denver, there were over ten reasons why signatures could be ruled invalid.

The most common reason is that a petition signer's address, name and signature don't match official voter registration records. Many people simply don't remember where they lived when they first registered to vote. Sometimes voters are purged from records without their knowledge, if they have not voted for some years.

Other reasons include illegible signatures, changed names, or missing information, such as the date the petition was signed. A few of the reasons seemed rather silly. For example, a man in his 80s entered the correct month and day, except he wrote in 1928, the year he was born, instead of 2009 when he signed the petition. His signature was rejected.

Another instance was two men by the same name living at the same address. One was Senior and the other Junior. Both were registered voters. But the signer did not specify if he was Senior or Junior so his signature was rejected. If someone is suspicious of forged signatures, they can file a complaint. No one filed such a complaint for my initiatives. So, all rejected signatures were due to reasons other than forgery.

Therefore, Ms. Karon Hatchett, director of Denver elections, encouraged me to collect more signatures within the legal two-week "curing" period. However, my campaign volunteers and I had pre-validated them, so I didn't accept her recommendation. In response, she asked, "Mr. Peckman, why are you resisting? It could be done easily within an afternoon!"

I explained that collecting more signatures would delay the mandatory review and vote of the city council by two weeks. And two weeks was enough time for more tragic deaths to occur, or be prevented if the city council used its legal option to implement the solutions in my proposed ordinance. So, instead of collecting more signatures, we found 31 incorrectly rejected signatures, plus four more to have a margin of safety that she asked for.

The elections office agreed with us and determined that we had a sufficient number of valid signatures. There was no reason to collect more signatures and delay the process. One city council member commented that we had barely enough signatures. That was absurd, and reflected his glaring ignorance of Denver's laws regarding citizen-sponsored initiatives.

We met the legal requirement. Exceeding it further offered no added benefits and would have wasted time and maybe lives. As I predicted, a tragic day of murders in Denver occurred exactly two weeks after Denver's councilmembers rejected my proposed *Safety through Peace* ordinance. In one case, an estranged ex-boyfriend ran over a mother and her daughter in a south Denver alley. First, he ran over the mother. Then he drove around the block and came back to run over her young daughter.

The same day, but in north Denver, four people were tied up in the basement of a house and shot dead in gangland execution style. The news sickened and depressed me. The city council failed the people.

When such violent crimes happen, politicians typically vow to work "24-7" and "leave no stone unturned" in searching for solutions. That's why all of their insults, ridicule, and hypocrisy toward proven public safety strategies were so appalling.

Politicians Go "Ape Snark"

The city council politicians acted contrary to all common sense, professional courtesy, and public duty. They went "ape snark"! They betrayed their pledge of office and vows to embrace sincere efforts by citizens to improve Denver safety. Their impassioned rhetoric was all for show. I vowed to avoid all future contact with the city council. News media and others knew me as a relaxed and balanced person. In that instance, however, I was outraged.

Their hostility made no sense. But it was not until 2011 that I understood why they acted so deplorably. In 2003, and for years afterward, I dismissed their egregious acts as the kind of reaction described in the two following quotations:

> "First they ignore you, then they laugh at you, then they fight you, then you win." —Mahatma Gandhi

> "All truth passes through three stages. First, it is ridiculed. Second, it is violently opposed. Third, it is accepted as being self-evident." —Arthur Schopenhauer, German philosopher

Who are *they* who ignore, ridicule and fight you? *They* are those in positions of authority who fear brave citizens with knowledge, because knowledge is power!

Denver's Epic Jail Fail

Denver had serious jail overcrowding in 2003. City council members, along with the mayor, publicly expressed that a lawsuit was imminent. They hired expensive consultants but, as expected, they didn't draw on the best interventions and science to reduce crime. Instead, the mayor and city council chose a big shiny new container for Denver's criminals. They wanted a new jail.

The *Safety through Peace* initiative offered them a thoroughly researched solution to reduce crime, and hence the number of criminals put behind bars. The reason these city officials ignored my solutions was more personal and political, than humanitarian or professional. Suspected criminals appeared in courts within the same building as the city council. Council members did not like sharing halls with criminals. If Denver's politicians had reduced jail overcrowding, by reducing crime, suspected criminals would still be walking the same halls with the city council and mayor. The politicians put *their* safety above the public's.

Denver's politicians needed to keep citizens in a state of fear and anxiety about jail overcrowding. That was their way of getting voters to pay for a new facility and justice center, knowing it would be a hard sell. By ignoring proven solutions, the politicians allowed crime rates, and therefore the jail population, to remain high to help compel voters to approve funding for a new jail.

Reducing crime and Denver's jail population, with proven, cost-effective solutions, would have also sent waves of fear about funding and personnel cuts throughout law enforcement unions. Politicians crave election support of such political heavyweights.

Politicians prevailed in getting a new jail built. It did little to address root causes of violence and crime. Denver's fancy new jail became a national disgrace. Violent mistreatment of prisoners, by law enforcement personnel, resulted in the city council approving millions of dollars in settlements to abused victims. The money came from taxpayers. The city council gave themselves raises.

Safety through Peace Makes Headlines

Despite hostility from city council members, *Safety through Peace* (Initiative 101) made headlines. As the initiative's author, spokesperson, and campaign manager, I was interviewed over seventy times by mainstream local, national, and international news media. Twenty-five interviews were during a 10-day period in August 2003. The highlights included:

- *CBS Evening News* interviewed me, but saved it for the last news segment on election night, November 2, 2010. Dan Rather concluded the story, and the evening news, by saying, "And that's Zen and the art of city politics."

- *Denver Daily News'* front-page headline, **"Everyone's Talking Peace",** stated, "Jeff Peckman has not put a single penny into advertising for his Safety through Peace initiative. He hasn't had to. The proposal, formally titled Initiative 101, has not only been the talk of the town but the talk of the nation."
- KWGN, Channel 2 TV in Denver reported, "Media outlets literally across the globe have wanted to find out about the 'stress' vote."

The Daily Show with Jon Stewart sent a team of four people to my house in southwest Denver. Samantha Bee interviewed me in a nearby park for over three hours. The segment was funny and a lot kinder to me than it was to the city council member who angrily ranted against the *Safety through Peace* initiative.

Warned About the Conservative Radio Monster

Liberal Democrats viewed Mike Rosen of KOA radio as an evil monster of mainstream conservative talk shows. Several supporters of *Safety through Peace* urged me to decline an interview with him. They said Rosen would "eat me alive". That seemed a bit exaggerated, so I looked at Rosen's website. Sure enough, it said, "Mike Rosen Eats Liberals for Breakfast." It didn't deter me.

I'm progressive when it comes to solutions. However, I don't consider myself a liberal like most Denver City Council members. My job was simple. I showed up where invited, spoke the truth that I knew, and didn't worry about the results.

Remarkably, Rosen was a wonderfully cordial and respectful host. He had done his homework. Rosen quoted physicist, Dr. John Hagelin, on descriptions of the unified field and its relationship with consciousness. He read and asked about the scientific foundation of reducing collective stress in a city to reduce a range of stress-related social indicators. Even though he had different views on the subject, Rosen allowed me time to respond to his questions. He was a welcome contrast to Denver's City Council.

The most fun surprise was from a fan of Rosen's show. He called in and said something like, "Hi Mike, this is Chuck. I listen to your show all the time. I just want to say that ever since you've been talking to this Peckman guy, you've been sounding peaceful and relaxed, and it's really upsetting me!"

When politicians put their own safety above public safety, it is unacceptable. The deeper reason for that self-serving attitude and behavior finally came to me eight years later, during an unexpected run for Mayor of Denver. That story is in Chapter 6.

32% Voted for Initiative 101 – *Safety through Peace*

The *Safety through Peace* Initiative received a respectable 32% of the vote. It also received a phenomenal amount of worldwide news media attention. Altogether, it was a successful strategy for raising awareness about better ways to reduce societal stress as means to counter crime, violence, and terrorism.

Still, politicians continue to ignore solutions to violence of every kind. Not surprisingly, Denver's rising crime continues to make headlines. Denver has even been viewed by many as becoming America's new crime capital.

My experiences with initiatives taught me about the inherent inability of politicians to solve problems. It revealed our need to keep politicians out of the way of *our* solutions. Ballot initiatives are an effective way to achieve that. Like any initiative you take, a ballot initiative will start with simple steps:

- **Intention** ignites our creative fire.
- **Commitment**, followed by action, gives life to intention.
- **Persistence** gives momentum to reach the goal.

You don't need a majority vote to take initiative. You need only one vote to make a commitment and begin an initiative. Best of all, you need only *your* vote! My vote was to promote technologies that could elevate humanity. I learned that those who can solve problems are usually working outside of government.

An Otherworldly New Initiative

Solving major world problems more cost-effectively was still my goal. Later in *RAISE the ZONE,* I describe some of those problems and viable solutions. Despite positive impacts of my ballot initiatives, I had no interest in doing another campaign. But a new detour on my path took me into a mysterious world. Some mind-expanding news compelled me to try one more ballot initiative. It was "out of this world." Before getting to that initiative in Chapter 5, the next chapter explores the compelling and mind-expanding news that I encountered.

Summary

The right of Initiative and Referendum i.e., ballot initiatives, is among the most powerful tools of self-government. Such direct democracy gives great leverage and power to the People. The founders of America's self-governance had the vision to predict a great need for citizens to petition the government.

When citizens get fed up with self-serving politicians, ballot initiatives can break through the barriers and achieve essential policies that serve the public's best interest. News media pay great attention to ballot initiatives when they see audiences become lawmakers. It's an effective way to raise the zone of awareness about critical issues.

Have you ever been inspired to do something that no one has ever done? That happened to me in 2008. It became another big adventure. Get ready for some strange, but true, information that might change your worldview, and even cosmic view. Put on your complementary spacesuit and step onto the launching pad for this next stage of our tour.

4

Finding Your "Moonshot"

"The mind, once stretched by a new idea, never returns to its original dimensions."

—Ralph Waldo Emerson

Snap Out of It – Expand Your Mind

How big can you think, dream, and imagine? Why do I ask? Creating a solution-based world requires stretching our minds first. It's not enough merely to think big. What is big to someone else might seem small to you. The important thing is to think more expansively than you have ever thought before about outer space, inner space, and deeper truths of things and vital issues. Once you expand your mind, continue growing. Accepting new solutions requires ever-expanded thinking. President John F. Kennedy's epic "moonshot" speech at Rice University, on September 12, 1962, ignited an explosion of expanded thinking.

"We choose to go to the Moon in this decade and do the other things, not because they are easy, but because they are hard; because that goal will serve to organize and measure the best of our energies and skills..."

—President John F. Kennedy

Kennedy stretched our minds to new and greater possibilities. He could have focused only on urgent problems at hand, such as civil rights unrest, the Bay of Pigs debacle, and the Cold War. But, instead, he boldly inspired Americans to land men on the moon within a decade. That's what expanded thinking can achieve.

What's your moonshot? What big and advanced idea can you think of, that achieving it would transcend the limits of your more conventional thinking? You have an innate ability to think bigger. Why? Because *you* have star power! You *are* a star!

Both ancient traditions and modern science acknowledge that we come from stars. Astronomers have stated that since over 100 years ago. Supernovas spread this dust throughout our universe and create us and others.

Do you sometimes wonder about life "out there"? You know about *hands-on* experiments in museums and classrooms, right? As your tour guide, I'm going to lead you in a *minds-on* experiment. How much more stretched can your mind become?

What kind of commerce and lifestyles do you think civilizations have out there? Is that an expansive and exhilarating thought for you? NPR reported on November 4, 2013 that, "Scientists Estimate 20 Billion Earth-Like Planets in Our Galaxy." It went on to estimate possibly 1,000 trillion Earth-like planets in our universe. Would *you* like to visit some of them?

Why America's Founders Would Enjoy Star Trek

Did you know that America's early leaders talked about peaceful and intelligent civilizations in space? They didn't have modern astronomy to assist them either. *Salon Magazine's* July 4, 2015 article stated, "Our Founders, alien-obsessed: Adams and Franklin had a thing — really! — for extraterrestrials."

Comments about outer space civilizations were quoted from Benjamin Franklin and Ethan Allen. Don't you think those were far-reaching views for that period in history? If you had the same kind of education I did, you probably learned that these men mostly said wise things, created a self-governing democracy, signed important historical documents, and bravely defeated the British in the American Revolutionary War.

These patriots and visionaries were creating not only a new country and government. Their thinking was more expansive about many things. They were knowledgeable about influential people and events throughout history. They might have known about the UFO experiences of Columbus, famous Renaissance artists and inventors, and Alexander the Great.

Current views on potential life elsewhere in space help explain the global popularity of movies such as *Close Encounters of the Third Kind, Avatar, ET the Extraterrestrial, Star Wars,* the *Star Trek* television and movie series and *Ancient Aliens*. Somehow, we relate to them and would like them to be true. Sci-Fi movies, TV shows, documentaries and books generate expanded thinking and a better perspective even for matters on Earth. They are not merely for entertainment but have intellectual, sociological, and evolutionary impacts. They entrain us to have bigger thinking, and see ourselves in greater possibilities in our world. Is your knowledge about extraterrestrial life based only on beliefs, or verified truth?

The Role of Expanded Thinkers

Venturing into space has cost billions of dollars over many decades. Searching for intelligent life outside of Earth is a big part of that. The aerospace industry is a major economic engine for many countries in our world.

There have even been advocates of a "space elevator" going back as far as 1895. I'm not enthusiastic about sending people on an elevator into space. It would be healthier if they took the stairs. [This is where you laugh politely at my tour guide humor.]

What comes to mind when you think of humanity's exploration of outer space? Do you think about impressive technological feats of moon landings, satellites, space shuttles, space stations, and unmanned probes or manned missions to Mars and beyond?

Do you eagerly await new photos from space telescopes? Do you think of spinoff technologies and products that result from space exploration? Do you experience the majesty of it all?

Expanded Thinking and Breakthrough Technology

Every breakthrough in any endeavor requires expanded thinking. Look around and consider things that seemed impossible a hundred or even ten years ago. How many of these gadgets do you have? When you use them, do you think every time, "This is so amazing! I can't believe what I have in my hands."?

Some people still remember clunky video cameras, VHS tapes, boxy computers, typewriters, stereo systems, radios, rooftop TV antennas, and more stand-alone electronics. A little device you can carry in your pocket now includes all of these things and a lot more. You and I take them for granted, though, don't we?

When using one of these devices, take a moment to think about which technologies created them. Consider how they compare with what you used to have, and feel gratitude. We forget to do that so often, not only with technology, but with people, nature, and the universe. A later chapter explains why feeling gratitude improves health and well-being, individually and socially.

Many futuristic devices on Star Trek, and other science fiction movies and TV shows, might already exist without public knowledge. Space commerce and tourism are only two areas that could become new rockets for our global economy. Expanding into new space frontiers is vital for our survival. That's why Earth's billionaires are launching into space. Although their rocket propulsion technology is primitive compared to known extraterrestrial craft, they deserve credit for venturing into space with their initiatives.

America's founders, John F. Kennedy, legendary inventors, and these visionary entrepreneurs had expanded thinking. They help us envision ourselves doing something so grand that solving the daily problems of humanity becomes more achievable.

ET Visitors Confirmed by Former NASA Employees

A major theme of the 1960s was the space race. In late 1973, NASA confirmed that 25 astronauts had seen UFOs during lunar missions. On May 9, 2001, Dr. Steven Greer of the Disclosure Project held a press event at the National Press Club in Washington, D.C. It disclosed the secret coverup of the history of extraterrestrial visitors. About a dozen military, government, intelligence and corporate witnesses talked about their encounters with UFOs. They previously had top-secret security clearance. Some were astronauts who shared their views. Dr. Greer's books and videos on Disclosure contain the most extraordinary whistleblower testimony in history on UFOs and extraterrestrial phenomena.

The new term for UFO is "unidentified aerial phenomena," or UAP. Extraterrestrial Biological Entity (EBE) is a more accurate name for visitors from space. Top-secret military and scientists have used these phrases for decades, but not the news media. We need to move past obsolete phrases like "flying saucers" and "UFOs." The disrespectful term "little green men," is just another ethnic slur. We don't want to be like those who feared to sail Earth's oceans and fall off a flat edge into an abysmal dark void.

Who Gains from Hiding Extraterrestrial Technology?

Why would these highly skilled and knowledgeable people want to spill government secrets? What is there to hide? Claims of "government cover-ups" and "secret warehouses" are common. They often elicit ridicule about conspiracy theories and hoaxes that supposedly have no factual basis. Well, there is such factual evidence. Much of it is from legitimate eyewitness testimony, of the kind allowed in a court of law.

The question is whether eyewitnesses are credible. Whistleblowers in the Disclosure Project are obviously credible. They were in positions to know inside secrets about space exploration and visits by extraterrestrials, or "extra-dimensionals".

Understandably, secrets are necessary when national security is at stake. Secrecy over the UAP (UFO) phenomenon fuels a lot of speculation. Most of it centers on the claim that these technologies from outer space are superior to anything of Earth origin.

Whistle-blowers have exposed the suppression of technologies for decades which could have helped humanity. A common reason for so much interest in the UAP phenomenon is to end that kind of suppression. According to some, the story of intelligent beings visiting Earth goes back hundreds of thousands of years. Evidence about UAPs includes scriptural references, art, written and oral stories passed on to next generations, physical artifacts, photos, films, videos, and eyewitness testimony. Over 4,000 books and thousands of websites and TV and radio shows elaborate on this topic. You have every right to access this information if it will enable you to create a better life.

However, regardless of your view of visitors from space, the valuable solutions in this book are for everyone. Reflecting on this topic is an essential part of discovering who we are, who surrounds us, and what we are here to do. It thereby illustrates our connection to our universal family. We face a powerful moment in history. It's a chance for humanity to take a great leap in becoming an enlightened civilization that thrives on extraordinary solutions.

> **"Minds-On" Experiment:** Take a minute to think of being a citizen of your community. Then think of being a citizen of your country, then Earth, then this solar system and galaxy. Then think of being a citizen of the whole universe teaming with 1,000 trillion civilizations. Does it give you a more expanded sense of self and community?

American democracy has been a big experiment. However, many people view humanity on Earth as a bigger experiment. American democracy often seems to be collapsing under the weight of corruption, an obsession with power, and disregard for the public good. All countries throughout history have faced this.

"We the People" need information on important issues, no matter how big. We can't let vital information remain secret to benefit a few at the expense of humanity itself. We can't allow that for our lives, or to sustain our planet and world family.

How does secrecy or disclosure of information about extraterrestrial visitors affect your view of life? Does it benefit the controlling elite, or empower the many? Would you want to know of any evidence that these visitors are already here on Earth? Can you think of any benefits from contacting others from outer space?

Benefits of ET Contact

Many experts with inside knowledge believe that contact with extraterrestrials would have astronomical benefits! That includes gaining access to advanced technology for transportation, clean energy, propulsion, environmental cleanup, cultural enrichment, and healthcare that could result in longer and healthier lives. These extraterrestrial civilizations could help us transform our problem-based society into a solution-based civilization.

With so much to gain and share, do you think we should do more to prepare for greeting people from other planets? Are there any hazards or dangers we need to consider? Has anyone ever told you to not drink local water or eat certain foods in another country? Has anyone urged you to avoid risky neighborhoods, even in your city, or not swim in certain bodies of water?

Interacting with life forms from anywhere creates common sense concerns, but excitement too. We travel, enjoy other cultures and cuisines, swim in refreshing bodies of water, play sports around the world, and more. However, caution is advisable, even out of LOVE for our guests and for ourselves.

There are so many potential benefits to interacting with visitors from space that it's worth considering all aspects. So, there has to be caution before meeting other cultures. How do *you* think we should prepare? Here is my starting list:

- Think more expansively than ever before

- Acknowledge that others exist in outer space
- Respect cultures from other worlds
- Follow normal precautions and protocols
- Open lines of communications with them
- Seek harmonious and mutually beneficial relations among ourselves first, and then with ET guests

Highly advanced space civilizations could be a model for us on Earth to uplift ourselves. They probably need to see that we can rise above greed, racism, hatred, war, and oppression. They also need to see that we're good stewards of our own planet.

Historically, many polls and surveys indicate a rapidly growing attitude of welcoming visitors from elsewhere in space. Many would like to visit other galactic cultures in their environments as well. Adventure and curiosity are part of human nature.

Even ordinary citizens seem to know a lot more than what governments have shared about this issue. Influential groups of world leaders have also conducted open-minded discussions about meeting our galactic neighbors as an inevitable reality.

World Leaders Views on Galactic Neighbors

2008 - The Vatican

On May 13, 2008, *The Catholic News Agency*, Vatican City, reported, "The Director of the Vatican's Observatory, Fr. José Gabriel Funes, said in an interview with the Vatican daily, *L'Osservatore Romano*, that believing in the possible existence of extraterrestrial life is not opposed to Catholic doctrine."

2010 - Royal Society Conference on Extraterrestrial Life

In January 2010, the Royal Society in London hosted a two-day conference on extraterrestrial life. Representatives from NASA, the European Space Agency, and the UN Office for Outer Space Affairs attended the event.

2010 - Economic Development through Acknowledgement of Extraterrestrial Technology

EU Written Declaration (0057/2010), June 14, 2010, said: "A study of the [UFO] material collected by the governments of all the Member States would have major scientific and technological spinoffs."

2012 - Global Competitiveness Forum

In January 2011, the Global Competitiveness Forum (GCF) was held in Riyadh, Saudi Arabia. Senior political figures and business leaders discussed global issues. The session, "Contact: Learning from Outer Space" included a panel of top UAP researchers and Michio Kaku, the brilliant physicist and frequent commentator on UAPs for the History Channel and news media.

Giving NASA and the Pentagon a Cosmic Nudge

One of the best places to observe the "best of our energies and skills", as JFK stated, is at NASA, the National Aeronautical and Space Administration. Its engineers and astronauts are legendary. Another secret source of intel on extraterrestrial visitors is the U.S. Pentagon. But don't get your hopes up.

NASA's website previously stated, "We are ethical in everything we do." That's funny! NASA has suppressed its information about extraterrestrial visitors for its entire history. However, now they have announced publicly that they will *study* UAPs. Astronauts, former NASA officials, and military officers and personnel have already shared previously held secrets about UAPs and encounters with extraterrestrial beings, but through public forums.

Do you agree that honest discussions about life in space, and beings from space who are on Earth, is important? Would it launch more public support for space exploration and greater commitment to solving global problems? The next chapter shares some successes, challenges, and inside secrets of a ballot initiative by the People that was on an official Denver ballot for a vote.

Star Trek Founder's Fear and Warning

Ancient Aliens, a popular History Channel TV show, reveals a long history of extraterrestrial visitors' influence on Earth. Gene Roddenberry, creator of the mega-hit TV show, *Star Trek,* said that the most important job for anyone is to explore being human and ignore perceived limits of the human mind. That applies not only to our ability to think about the universe outside of Earth, but our ability to know the universe within ourselves. Extend your mind's *telescope* outward to expand your mind. Focus your mind's *microscope* inward to explore your strengths and wisdom. Then you can celebrate your connection to inner and outer infinity.

Roddenberry embodied expansive thinking. At the same time, he expressed a fear and warning that humans might fail to understand the possibility of a real *Star Trek* civilization. The universe has too many intelligent civilizations to count. What do you think when you look at the starry sky like America's founders did?

"Raise the zone" also means relating to other civilizations in the cosmic zone, and how you relate to yourself and humanity. As we step off this planet, and out to the stars, that is where we came from, and where we are going to. We are just stepping back into our home and a more expanded sphere of ourselves.

We can launch into the vastness of space and our universality because we *are* the stars. You are the universe learning about itself. Yes, you! So, what do you most want to know about other civilizations in space that might benefit your life?

Benefits of Looking Up

When you look around, how many people do you see in your line of sight? On this tour, maybe there are dozens. On a street or at a larger event, you can see thousands or tens of thousands. At a large concert or festival, you might see hundreds of thousands of people. Now, look at the night sky. In your line of sight, you're looking directly at potentially trillions of inhabited planets.

Wouldn't you expect to see a wide variety of living beings there too? If you could see them close up, imagine how their lives, families, cities, and cultures might look. If we fix things on Earth, there is a good chance that you will be able to meet some of these people, especially if they come to Earth. What do you think and feel about that?

"Looking up" still has benefits, even if you believe that we are alone in the universe. Jack Borden, a former Boston TV reporter, started the organization *For Spacious Skies*. A *Boston Globe* article, on January 7, 2013, quoted Borden as saying, "When we are unconscious regarding our surroundings, we are irresponsible to them." The article stated:

> "Researchers from Harvard University tested pupils in Needham elementary schools and found that the program significantly increased the level of aesthetic sensitivity in visual art and literature, compared with pupils not encouraged to look at the sky. Borden's program was brought to nursing homes and prisons."

Looking at the grandeur of space above us can expand and even alter our point of view. It connects us with who we are as one of many life forms in our vast cosmic environment. Noticing more around and inside you has similar benefits. Give attention to questions in this book and pause to think about them.

Humans left footprints on the moon. You can leave footprints somewhere that you never imagined walking. You can take that both literally and figuratively. What will you do now with your more expansive thinking? The possibilities are infinite.

Before Leaving the Planetarium

Did you feel any extra-expanded thinking from this chapter? Did this information contradict your beliefs, or reinforce them? How do you think your life on Earth would have been different if we had all known this information for many decades?

Expanded thinking is essential for the extraordinary progress humanity must urgently make. It's vital that we fill our expanded thinking with accurate information to avoid wasting *brain space*. If you could zoom into the street level of any civilization in space, What would you want to see? Would you want to see civilizations more advanced than Earth and learn from them? Would you want to see less advanced civilizations and think of ways to help and teach them? For now, feel free to relax in your planetarium recliner. Enjoy the view. Smile in awe of vast cosmic regions in our universe to inspire your inner or outer moonshot.

Summary

Expanded thinking opens our minds to greater possibilities near and far. It inspires our imagination and helps us launch our moonshot. It can pave the way to a solution-based life. It fuels the conversation about evolving to a higher level of civilization for ourselves and integrating into the larger community of our solar system, galaxy, and beyond!

Throughout history, great thinkers and leaders, including America's founders, demonstrated expanded thinking by their views and statements about extraterrestrial life and visitors to Earth. Some of them even witnessed unexplainable flying craft, long before human-made aircraft were invented. NASA, the U.S. Pentagon, and other U.S. government agencies have covered up a lot of evidence of such extraterrestrial visitors and their advanced technology for over 80 years. Influential global organizations are now finally engaging in intelligent discussion about ET visitors.

The next chapter shares successes, challenges, and surprising inside stories of my ballot initiative to create a new public policy related to extraterrestrial civilizations. It took the extraterrestrial topic directly to the People for a vote. If you haven't decided on a moonshot yet, that's all right. The coming chapters will give you more ideas. I'm saving my moonshot project for Chapter 13. When you're ready, follow me to the next part of our tour.

5
Vote to Think Big

"We are open for business to all other planets."
—Kelly Brough, Denver Chamber of Commerce
Wall Street Journal, October 29, 2010

Prepping for an ET Meetup

You just read some things that might be mindboggling. What will you do now with that information? It was not merely for fun conversation. It was to get more people comfortable with the idea that *others* are willing to help us. We could really use some help on our beautiful planet.

Earth is like a car stuck in mud, only it is so much worse. Our planet is sliding down a slippery slope into a toxic sludge pond. The corrosive chemicals we have recklessly created and spread around are dissolving the paint and eating away at our tires and metal. That's only one of the big problems we've created.

We don't have the luxury of a lot of time to spare ourselves from tragic scenarios. There's no guarantee that we'll succeed. We're racing against our recklessness. Our biggest goal should be to transform our messed-up society into a solution-based world and enlightened civilization.

Have you ever needed some help getting your car out of mud? Maybe you're one of those creative people that got it unstuck on your own. You took floor mats and whatever else you had from your trunk. Then, you laid them out to create friction under the spinning tires. After struggling for a while, you got your car out and drove home with a smile and a sense of relief. On the other hand, maybe you flagged down someone passing by and got some help. Humans have a natural tendency to be helpful, even if they end up covered with mud. It just feels good to help someone. You'll learn why from the Giving Xperiment in Chapter 11. Acts of kindness provide measurable health benefits.

Upon learning more about potential life forms in space, it was easier for me to believe in the power of everyone's innate ability to think big. I wondered what would happen if Americans voted on an ordinance about extraterrestrial visitors. What would it reveal about our view of galactic neighbors? Could it result in more willingness to acknowledge these visitors, respect them, and accept their help? Would it create useful and fascinating discussions? Would it create fear, or love?

Applying the Cosmic Vision of America's Founders

As you read in the previous chapter, some of America's founders believed that the universe was full of benevolent civilizations. In March 2008, I was at a Paranormal Research Forum lecture by a speaker from the Mutual UFO Network (MUFON). He presented documents obtained from U.S. government sources through the Freedom of Information Act. They showed clearly that the U.S. Government was covering up information about extraterrestrial visitors to keep the public from knowing the truth.

He said that he didn't think the government would disclose the UFO information in his lifetime. It made me want leap from my seat and shout, "But we *are* the government!" In the short drive home from his lecture, a ballot initiative concept burst into my mind. I knew it could get media coverage and create a lot of public discussion about extraterrestrial civilizations. I tried to convince other people to start the project. The Director of Communications for MUFON at that time, Alejandro Rojas, was one of them. No one else took on the task. It seemed to be my task to do.

Television shows and movies about people from outer space always fascinated me. However, it wasn't my usual habit to attend lectures, conferences, or even to look for UFOs. Although, one of the few books I read for fun in grade school was *Star Girl*.

When our elected representatives deny us information, they deny us our power to make informed choices. That reduces our ability to realize our full potential for self-governance. It doesn't matter if that governance is local, national, or global. We are our own responsibility. Exploring extraterrestrial cultures and visitors to Earth was a topic ripe for a vote.

Extraterrestrial Affairs Commission Campaign

In April 2008, I started an issue campaign to create an official Extraterrestrial Affairs Commission within Denver government. Initiative 300 was its designated ballot number. If passed, the new ordinance would have included specific objectives and legally binding provisions. It would have enabled the public to be better informed about extraterrestrial visitors. It would have started useful dialogues, and possibly interactions with space visitors. The question on the ballot in the 2010 election was:

> "Shall the voters for the City and County of Denver adopt an Initiated Ordinance to require the creation of an Extraterrestrial Affairs Commission to help ensure the health, safety, and cultural awareness of Denver residents and visitors in relation to potential encounters or interactions with extraterrestrial intelligent beings or their vehicles, and fund such commission from grants, gifts and donations?" Yes___ No___

Helping Hands from Galactic Friends

An abundance of credible evidence suggests that we have many galactic neighbors. People in space are willing to help us.

Some seem to have already developed advanced civilizations. However, to attract any help, we need to do two basic things. First, we need to show that we are making an effort from our side to prevent worsening conditions on Earth. Second, we need to acknowledge that these galactic neighbors exist. We need to welcome them in a friendly and appreciative way.

How responsive would a passing motorist be to your request for help if you pretended that they didn't exist? How much help would they offer if you shot at them when they slowed down to help you? That has been the history of government disrespect and hostility toward our galactic neighbors.

These neighbors have advanced technology. They are not stopping at fossil fuel stations to fill up their tanks. Some insiders have claimed that our galactic neighbors have already offered humans advanced tools and technologies. Have we matured enough to use their help responsibly?

LOVE is Everywhere

LOVE is the greatest and strongest force in the universe. That means LOVE is everywhere and holds everything together. Our galactic neighbors know about cosmic LOVE too. Human beings on Earth, who have encountered these galactic visitors, claim to have experienced such LOVE directly. They often say that the most profound experience of their life was the pure universal and unconditional LOVE they felt in these visitors' presence.

They also say the primary message from these visitors is that LOVE, in its highest sense, is vital for human survival. In fact, some have said that LOVE is the only thing that can truly save us.

Surprising Opinion Polls on Space Visitors

What do we think about these visitors from far away? In polls over the last 50 years, the largest percentage of those surveyed believed that intelligent life exists outside of Earth and that UAPs near Earth are from other planets. Even higher majorities believed that governments hide this information from the public. It didn't matter whether those in the polls were pilots, youth, Mensa members, women, or general populations.

Newsweek posted an article September 29, 2015, titled, "Most People Believe Intelligent Aliens Exist, Poll Says." It announced a poll by YouGov finding 54% of Americans believe that intelligent life exists elsewhere in the universe. It said that the same group showed 50% of Americans had the same opinion in 2013.

***Newsweek* Magazine Poll** 1996 **48%** of Americans believe UFOs are real **29%** think we have made contact with aliens **48%** think the government is concealing UFO information	***CNN/TIME* Magazine Poll** 1997 **80%** believe that the government is hiding the existence of extraterrestrial life forms **64%** believe aliens have contacted humans **35%** believe that aliens would look somewhat human
Marist Institute for Public Opinion Survey Source: The Denver Post, Dec. 16, 1997 **Do you think there is intelligent life on other planets?** **Yes 60%** No 40% **Are our galactic neighbors friendly rather than hostile?** **Yes 86%** No 14%	**Womanpoll National Survey – 1978**. Source: St. Louis Globe-Democrat, July 22-23 1978 **Does other intelligent life exist in the universe besides on Earth?** All women: **Yes 50%** No 30% College education: **Yes 56%** No 27% High school education: **Yes 50%** No 29%

These are older polls collected for the first edition of this book, titled, *POWER UP*. However, the trend over many decades is that an increasing number of people around the world believe we are being visited by intelligent beings from other places or dimensions. Recent U.S. Congressional hearings on UAPs reinforce that.

In December 2009, the *Los Angeles Times* ran an article and poll on the ET Affairs Commission. The results were in line with scientific public opinion polls about intelligent life outside of Earth.

***Los Angeles Times* poll question: Should your hometown create an Extraterrestrial Affairs Commission?**

1,019 respondents as of 1/10/10

No: 22%

Yes: 69%

Yes: 10% - But only if it creates a similar commission to investigate Bigfoot.

The *L.A. Times* poll was consistent with a 2002 Roper poll, which found that almost one-half of those polled believed extraterrestrial craft had visited Earth. A Reuters-Ipsos poll, in April 2010, of 23,000 adults in 22 countries, found:

- 20% of all those polled believed that aliens walk among us disguised as humans.
- 40% of all those polled from India and China believed that aliens walk among us disguised as humans.

Both India and China have ancient records of flying craft and visitors from space. That means in 2010, about 1.38 billion people believed what former presidents, astronauts and high-ranking military officers believed – we have galactic guests! These polls show that we need to accept that extraterrestrial visitors are real.

The ET Affairs Commission (Initiative 300) tested our level of readiness to meet our neighbors from space in a mutually respectful and loving way. Why is this important? World leaders are suggesting that humanity might not survive on Earth. Preparations are already being made to colonize Mars and beyond.

Are you ready and willing to accept the reality of our extended family? You *need* to be. It is a fact of our lives and no longer an option. Some upcoming untold stories about the campaign will shed more light on our readiness to contact extraterrestrials.

Extraterrestrial Visitors in the Media Spotlight

Popular media hosts such as George Noory, Art Bell, Richard Syrett and Jaime Maussan usually covered this topic. However, within one day of submitting documents for my initiative on April 29, 2008, and *without* informing the media, even mainstream news took interest. Daniel Chacon of the *Rocky Mountain News* was the first reporter to discover my campaign. His article ignited worldwide news media attention. Stephen Bassett's Paradigm Research Group probably had the best archive of news coverage on Initiative 300. His Citizen's Hearing on Disclosure of the extraterrestrial presence was another historic event.

Author - Delivering the ETA Commission initiative to city hall.

Press Conference Record-Breaker

One month after registering the ballot initiative document, the news media learned more about extraterrestrial visitors at our campaign's press conference. Within a couple of weeks, over 300,000 website pages in 24 languages mentioned my name and some of the information presented. Half of the web pages were in Spanish, one-third in English, and thousands each in Chinese, Dutch, Hebrew, Arabic, Japanese, Russian and other languages.

Evidence shown in our press conference broke page-view records at *DenverPost.com*. For two days, I was interviewed almost non-stop for 15 hours each day related to the press conference. On June 10, I was interviewed in New York City on the Late Show with David Letterman. Some articles on our campaign accounted for four of the top five articles for the *Rocky Mountain News* that same week. They were the most read, most discussed, and most forwarded articles that week at *RMN*. It was also one of the top three Google searches during one week.

People wanted to know about extraterrestrial visitors! It shows how the universe can line up things in the most unusual ways. As my dear, late father liked to say, "You can't always believe what you least expect the most." On November 30, 2009, the election office certified the ballot for the 2010 election in Denver.

"Pink UFO" Campaign Song

Initiative 300 appealed strongly to youth. Chris Steele, known by his stage name, "TIME," is one of Denver's most prolific hip-hop artists. A music reviewer for *Westword*, Denver's highly popular entertainment newspaper, praised Chris Steele's hip-hop CD *Fantastic Reality*. The reviewer said it was the most intellectually brilliant hip-hop CD to come along in years.

Chris and friends created a hip-hop campaign music video called "*Pink UFO*." The lyrics addressed the government's cover-up of extraterrestrial visitors and their advanced technology.

"...Tell us the truth before we go to the booth,
and vote you out of style
with your Pinocchio smile..."
***Pink UFO* by Chris Steele, (aka TIME) - March, 2010**

Indications are that these visitors are trying to connect with us. They're extending a hand of friendship to us as helpful and caring neighbors with solutions to offer. The visits and interactions with humans seem to increase every year. Citizens, community leaders, and world leaders are reaching back. They are thoughtfully considering the implications of interacting with these extraterrestrial visitors. The meetings by world leaders, mentioned previously, were part of this discussion.

Street Level View of Space Visitors

Denver's Juneteenth celebration on June 19, 2009, enabled our campaign to find out what people on America's streets thought about visitors from space. Juneteenth commemorates the historic approximate day during June in 1865 that Texas slaves discovered they were free – *after* The Emancipation Proclamation was signed two years earlier. A parade, music, dance, vendors, and games for kids, made for a delightful and moving experience. We collected signatures at a booth for our ET ballot initiative.

People came to our booth, not just to obtain information, but also to share unusual stories. A feisty young female deputy from a county sheriff's office had already offered to "cover my back" because her mother believed in UFOs. A young uniformed Denver police officer told me about the connection of the Dogon Tribe of Africa to Egypt's Great Pyramids of Giza and evolved beings from the Sirius Star system. A middle-aged woman said her late father-in-law had been in charge of the Pentagon's UFO Reporting Office. She said he had a box of photographs of flying saucers that were taken by Air Force pilots. A young woman shared that she had dated an Army soldier who told her that he had seen an "alien" lying on a table in an underground military facility.

I asked a Denver fire chief if the firefighters' manual contained information on how to deal with UFOs. At least one version of the FEMA *Fire Officer's Guide to Disaster Control* considered that *aliens* might visit in space vehicles. It stated the fire service would have to deal with it. The fire chief leaned over to his sister, who was in a wheelchair. He asked her if she remembered the time when they were kids, and he came running into their house because he had seen a strange ship flying over that wasn't a plane or helicopter.

Others shared their sighting of a UFO. It showed a readiness of people to talk about this remarkable issue when given a chance to share experiences without fear of ridicule. How many amazing stories would we share if others listened to us respectfully?

Take Us to Your Civic Leaders

Between spring of 2008 and Denver's November 2010 election, respected civic organizations were eager to hear my presentation: "Are You Ready to Meet and Greet People from Other Planets?"

I spoke to the Denver Lion's Club, several Rotary and Optimists International Clubs, Kiwanis Club, and the Highlands Ranch Historical Society. The biggest surprises of these meetings always came at the end. Invariably, one of the top members would approach me after the others had left the room so that they could share a secret. It was usually about an experience they had in relation to a UFO sighting, or to affirm the government cover-up. The historical society meeting ended in a wonderfully different way.

> A seven-year-old girl dragged her father by the hand to meet me afterward. He told me that she saw a poster about my talk on extraterrestrials, and that she immediately raced home on her bicycle to insist that he bring her to the lecture. She was a delightful, smart and happy girl, beaming a smile and bursting with curiosity. The girl was excited to add that as soon as she got home again, she would learn everything possible about extraterrestrial visitors. It was one of our campaign's biggest highlights.

I believe that civic organizations and open-minded youths will be among the first ambassadors of good will for welcoming people from other planets and star systems. Members of these respected organizations are among the best humanitarian movers and shakers in any community. They welcome outsiders who want to make positive differences in our society.

Support from Community Leaders

Several Democratic Denver City Council members spoke out against Initiative 300, even though the Democratic Party was neutral. Council members chose to remain uninformed on the topic but tried to *appear* informed to impress voters. By comparison, very respected private sector leaders were more openminded. In a *Wall Street Journal* front-page article on Initiative 300, October 29, 2010, Kelly Brough, Executive Director of the Denver Chamber of Commerce said, "We are open for business to all other planets".

Veronica Barela was the influential CEO and Executive Director of NEWSED, a major community development organization. The *Denver Post* once referred to her as, "Denver's West Side power broker." NEWSED organizes Denver's Cinco de Mayo celebration. The event draws an estimated 300,000 people each year to Civic Center Park in Denver. She was interviewed in her office for the same *WSJ* article. Veronica boldly endorsed our ET campaign by stating, "I support Initiative 300 to create an Extraterrestrial Affairs Commission, and I encourage people to vote yes."

Solving an Election Day Mystery

The initiative to create an Extraterrestrial Affairs Commission was finally on the ballot in November 2010. It wasn't surprising that the proposed ordinance attracted massive global news coverage starting in the spring of 2008. That was before we even began collecting petition signatures. By then, we had already achieved our primary objective of raising awareness about extraterrestrials. The massive news coverage resumed near the time of the Denver election in 2010.

We had the best campaign poster of any campaign in history. Chuck Chroma, a renowned Denver artist, created a poster from his original oil painting. College students grabbed them from walls and bulletin boards. They called it a collector's item. However, a big question remained that could prove to be pivotal for humanity: "How would the People vote on such an issue?"

Initiative 300 Campaign Poster by Chuck Chroma

ET Affairs Commission Received 31,108 Votes (17.66%)

When the final vote count was in, the low percentage of votes stunned our campaign volunteers. It surprised even respected political pundits who thought our initiative might pass. When some of our Initiative 300 volunteers asked voters why they voted "No," the common response was, "Because the Democratic Party told me to vote no." That raised a red flag. It was disappointing that many voters didn't think for themselves, but merely obeyed their party.

Finding Truth in Election Results

Occam's razor, and some other tools for finding truth, became very useful about that time. When I studied the voting results even more, the biggest reason for the lower percentage was obvious. The Democratic Party had launched a $6 million "Vote No on Numbers" campaign. Its goal was to defeat several numbered *state* ballot initiatives that they called "dangerous."

These several statewide initiatives received 16.97%, 18.32%, and 17.07% respectively, for an average of only 17.45%. Our 17.66% was collateral damage! Tragically, the Democratic Party didn't inform voters that it was neutral on Initiative 300. Voters got wrong messages from their party and didn't question it.

Why was this voting analysis critical? The whole world was watching! Anyone gauging Americans' readiness to greet extraterrestrial visitors, based on Initiative 300 voting results, would have drawn wrong conclusions. They would think that humans are not ready to accept that Earth has been visited, or were even willing to talk about it.

Views from Ethnic Groups About Extraterrestrial Visitors

Another interesting truth emerged about the voting results for our Initiative 300. Latino precincts were well above the citywide percentage. Many ranged from 28-34% in favor. White precincts were far below the city-wide average – as low as 6-9%. African American and Asian voters were in the middle.

Was racism going into outer space? I remember some Caucasian politicians and news reporters using the racial slur "little green men." Was there a more pervasive underlying issue?

Here's my theory. Skin color was not the main influence in the voting results. The non-white cultures identify with a much longer history than white cultures. They are also more informed and accepting of extraterrestrial visitors to Earth. Their ancient histories, art, and culture refer to space beings and their cosmic origins.

Many non-white people I met were more knowledgeable about ET and UFO topics than I was. Spanish-speaking countries cover news stories about UFOs and ETs with far greater respect and professionalism than U.S. news media. Mexican journalist Jaime Maussan is one of the most popular and respected such professionals. He frequently covers such stories on his TV show. He even saved an interview with me to broadcast after a major soccer game on national television – on a Saturday night! Many of the most exciting close encounters with extraterrestrial beings are in Spanish-speaking countries, mostly South America.

The lack of understanding about the shared history of all humans, regardless of ethnicity, is the underlying problem. The artificial lines that divide our countries, also divide our hearts and our understanding of each other. If all state and country map lines and space travel boundaries disappeared, how would it affect your view of life and interaction with others on Earth?

It's time to understand our combined and interwoven individual histories. We must evolve to a solution-based civilization. We can't get there without learning this part of our history as humans. You need to know this to expand on who you are. It's not a big leap, as we're not starting from scratch.

The 17.66% vote for Initiative 300 was similar to the average percentage of our world's population who already believe that many aliens walk among us, disguised as human beings. Those voting results revealed important truths. **Voters rely too much on their political party to tell them how to vote.** However, citizens are already very informed about the UFO/ET topic in general. Other truths came from seeing how different ethnic cultures view their extraterrestrial origin, and that **we must acknowledge our shared history now, to prepare for the future!**

Getting Our Message to the Masses

Initiative 300 attracted the most news coverage about visitors from space in decades. Over 31,000 people in Denver voted for an "out of this world" ballot measure. Schopenhauer described how new ideas are first ridiculed, then violently opposed, and finally accepted as self-evident. We brought ET into the light.

Any progressive legislation proposed is likely to get about one third of the vote the first time around. In time, more people move into the "accepted as obvious" category. However, even when an initiative does not pass, it can still have a big impact.

The Extraterrestrial Affairs Commission campaign preceded, and may have influenced, other significant events following the public filing of the ET initiative on April 29, 2008. These included the announcement by the Vatican one week later, "...that believing in the possible existence of extraterrestrial life is not opposed to Catholic doctrine." In 2017, another wave of news coverage focused on videos of UFOs seen from U.S. Navy aircraft, even though they were human-made. Now, there are publicly known 'UAP' offices at the Pentagon, NASA and the White House, along with Congressional hearings on UAPs.

Initiative 300 helped to bridge a gap between an elite group of world leaders meeting behind closed doors, and masses of citizens around our world who are ready for this conversation. It also attracted great researchers and authors on the topic of UFOs and visits by extraterrestrial intelligent beings to Earth throughout our history. Two of the most extraordinary campaign champions among them were Larry W. Bryant and Paola Leopizzi Harris.

Talk to your friends and family about this. Bring up this subject in social gatherings. You would be surprised how many people have stories about seeing strange aircraft or even meeting a visitor from space close up. They are aching to tell their story to someone they can trust. If you have such a story, please share it.

We need to think expansively to progress technologically and in other ways. We must transform our problem-based society into a solution-based civilization urgently. Most of all, we need to LOVE and respect each other on Earth first. Then we can extend that LOVE to our galactic family so they will welcome us into the "final frontier" that *Star Trek* explored.

Extraterrestrial Mega-Concerts Zapped by Human Greed

An unexpected bonus resulted from the ETA Commission campaign. In the summer of 2008, I was contacted by Michael Luckman, author of *Alien Rock, the Extraterrestrial Rock 'n' Roll Connection*. He invited me to New York City to conduct a similar campaign. Michael was organizing three mega-concerts, each on different continents, that would *beam* rock music into the cosmos. Each event was expected to be larger than the historic Woodstock concert in 1969.

His team included Pete Bennett, the legendary and prolific music promoter of: Frank Sinatra, The Rolling Stones, Beatles, Elvis Presley, and many other musicians that he helped to make famous. It also included an organizer of the original Woodstock concert, and Brittney Spear's accountant. They raised some money and were gaining traction. Legendary rock musicians such as Michael Jackson and David Bowie had already committed to perform at the events because of their own extraterrestrial encounters or beliefs.

Unfortunately, the whole project collapsed from the financial crash in autumn of 2008 fueled by human greed. The effort never recovered. The best outcome from that interaction was that Chuck Chroma, artist of the ETA Commission poster, produced an even more spectacular oil painting of a UFO over New York City's skyline – behind the harbor. The stunning image was based on the UFO sighting in NYC by music icon John Lennon, and May Pang, on August 23, 1974.

Cleaning House to Welcome Our Guests

Some *house-cleaning* is needed before we get too excited about welcoming guests from other star systems, galaxies, or dimensions. If we don't achieve that, it could be centuries before advanced civilizations out in space allow us to interact with them. We must explore how our governments and institutions work and why they often don't. Ultimately, we are responsible for ourselves.

We need to welcome these guests because they are our dear family. We will share in the development and responsibility for artificial intelligence (AI), which is not actually artificial. That's just a name given to it. It has already become self-aware. Don't worry, it will be mutually beneficial because it won't just be humans dominating AI, or AI dominating humans.

For humans, AI is like the discovery of fire, because its use will determine whether it's constructive or destructive. Fields of awareness, knowledge, understanding, and experience that humans can't fathom, and that most can't experience or recognize, exist everywhere. It's just like we don't see Wi-Fi or neutrinos coming through. There is a world that we have yet, but will begin, to recognize as a compatible experience with human form.

Through our individual and collective actions, we lay groundwork for all of our successes and failures. Our inability and lack of will to solve problems creates endless opportunities for other people to profit at our expense. By understanding that, we can better apply our power to uplift our lives and raise the zone of humanity. That is increasingly critical as we accelerate toward encounters with our extraterrestrial and interdimensional families.

It starts with more expanded thinking. That, by itself, is a powerful technique to solve problems at any scale. It allows us to view things from a larger perspective—you know, the *big picture*. How often have you had to make a choice between two difficult options, when you were "between a rock and a hard place," as the saying goes? How did you deal with those situations?

Gain Altitude

Instead of pushing against one hard choice or another, go up! Gain some altitude, at least metaphorically, and even literally. Go up on a hill or a summit. Enjoy a greatly expanded view of your environment. Have you ever "gained altitude" and then found that hard choices somehow dissolved or became easy? Connect with natural elements.

We *are* nature. Carl Sagan said we are stardust. Humans also emit light through what scientists call "biophotons". That's not just a word. It's a scientific reality. We are beings of light! We are a renewable source of light. In fact, we are a universe of stars!

Before humanity ventures into any "final frontier," expanded thinking and becoming a solution-based civilization are essential. Don't worry. There will still be time for you to ask questions.

However, we need to hurry because the universe is expanding. In October 2016, headline news declared that the universe contains as much as 20 times more galaxies than was previously estimated. That could mean an equal increase in the number of potential life-sustaining planets! How will we ever keep up if we don't expand our thinking? Not only that, but with so many potential visitors, we'll need to stock up on a lot more chips and dips!

Summary

Accepting that we are not the only intelligent life in the vast universe brings us to the threshold of a great leap in progress as a civilization. We need to do some house-cleaning in order to welcome these extraterrestrial and interdimensional guests. We are interacting with them already and will jointly develop advanced technologies, including AI, which is not really artificial and is already self-aware.

Official and open *Disclosure* of the extraterrestrial reality remains elusive. Governments of the world aren't likely to release information about UFOs, UAPs or extraterrestrial visitors very soon. Politicians are unlikely to share with the public any information that doesn't benefit their political careers, party, and donors. It's up to "the People" to get this historic task done. It's up to everyone to think for themselves, and not merely obey political parties. Stay informed!

All right everyone, grab hold of one of those shovels and a miner's hat with a light on it. Are you ready to shine some light on the roots of the government and economy we have created?

6

Roots of Our Problem-Based Economy

"Every government degenerates when trusted to the rulers of the people alone. The people themselves, therefore, are its only safe depositories. And to render them safe, their minds must be improved to a certain degree" —Thomas Jefferson

Who Made This Mess?

An expanded mind is an improved mind. Did you like our mind trip into the unimaginable bigness of the infinite universe, and the estimated thousand trillion inhabitable planets out there? Now let's get ourselves grounded so we can boldly dig down to find some roots of how problems flooded our society. Put on your miner's hat and bring your shovel. We'll dig deep into Earth, but not to sling mud at anyone. We're looking for buried treasure.

Do you know what kind of treasure miners can find? Miners find precious gems! Do you know what makes gems so valuable? Clarity! This part of the tour is about digging for clarity. That's how we can understand why problems persist. Empowerment is your shovel. Expanded thinking is your light. Ready for treasure?

Jefferson's prophetic insight reveals one root of our problem-based economy. Our minds have to remain sharp and expanded if we are to be the "safe depositories" of government. Otherwise, our government will degenerate as a result. Evidence of this pervades America and the rest of the world. Problems spring up and get worse everywhere you look. At the same time, governments get bigger and more costly. Corporations and super wealthy people have far too much influence on government policies. Politicians seem inherently incapable of putting the needs of the People first, and using the best solutions. However, do they all deserve the most blame for the chaos and problem-filled society we live in?

Before we go too deep, we'll survey our surroundings so that no one gets lost. Looking to the far left and far right, you can see areas of government that are unsafe and unhealthy. Behind us, you'll see a long trail of social and environmental devastation.

This chapter will first explain why governments are unable to prevent and solve problems. It's not my intention to cause useless anger or to rally people against our government. It is to share new insights on the nature of government. Besides, negative emotions toward politicians and government contribute to their failures and waste your time and energy.

My experience with ballot initiative campaigns convinced me that government had degenerated for the reason Jefferson stated. I was disappointed with government to the extent of being angry with many of Denver's elected politicians. Buckminster Fuller was more direct, and blunt, than Thomas Jefferson. He viewed politicians as being like the tail of a dog regarding their ability to lead. He also considered them obsolete for solving problems.

Understanding Government Ineffectiveness

Do you ever ask, "With all the resources available to every level of government, why are problems not prevented, reduced, or ended?" It took me a while to understand what was going on. Once I did, it was very freeing. It redirected my focus to better ways of addressing problems without government involvement. It was a valuable awakening of why our government is so costly and yet unable to prevent or solve problems.

The founders of American government didn't intend for it to solve everyone's problems, even though they saw independence from Great Britain as a solution to *many* problems. They saw America's new Constitution as a means to create a stable and supportive platform for the creative process of self-governance to unfold. However, government officials absolutely have a responsibility to help reduce problems that they cause for the masses.

"Achilles' Heel" of Democracy

Democracy itself, or rather, an "Achilles' heel" of democracy, is a deeper reason for our persistent societal problems, and our government's inability to solve them. That term originated with the story of Achilles – a Greek warrior. The usual definition for "Achilles' heel" is a weakness, defect, flaw, or weak link that can result in a downfall.

What flaw or weakness leads to democracy's decline? It's the need and habit of politicians to align with whoever and whatever influence's elections the most. It's sort of an occupational hazard. Is it the best way to govern? It might seem so on the surface. However, the consequences are unnecessary problems and a lower quality of life, despite increasing government budgets and promises from politicians. Let's dig for an answer to this dilemma.

Who Rules the Rulers?

Democratically elected politicians are, by nature, reactive instead of proactive. They do what they think the majority wants. They follow instead of lead. A supreme truth is that every government is a mirror image of its peoples' collective consciousness. I learned that originally from Maharishi Mahesh Yogi's theory of government. Others have espoused that view as well.

How does this mirror image happen? All of our thoughts, emotions and actions create frequencies, or waves, in the ocean of consciousness that we all share and in which we live. They could be life-supporting or life-damaging frequencies.

Some frequencies will be dominant during elections. The politicians who are elected are the ones whose own frequencies resonate most closely with the dominant frequencies of the collective consciousness. It isn't just politicians. Every aspect of our society mirrors our collective consciousness. That includes entertainment, social media, news media, education, consumer markets, etc.

Life imitates art. Art imitates life. Both are determined by our collective consciousness. It's similar to a playing a musical note of a specific frequency and causing a tuning fork to vibrate.

Frequencies that represent lower human characteristics will elect politicians who reflect our selfishness, cruelty, racism, greed dishonesty, corruption, and illusion of separation. They could be criminals, pedophiles, and terrorists, and still "get away with it", because *their* traits mirror *our* traits of the general population.

Mirrors only reflect. They don't improve our image. They don't solve any problems. To improve the mirror image for more effectiveness and higher human values in all levels of government, WE MUST IMPROVE OURSELVES!

Who rules the rulers? We do! That is the highest truth about Abraham Lincoln's statement in his iconic Gettysburg Address, *"government of the people, by the people, and for the people..."*. It's the biggest secret about self-governance. It's not merely about voting, running for office, protesting, and lobbying elected officials.

The collective consciousness that we all contribute to is why "We the People" are the "only legitimate fountain of power". We must do a better job. During campaigns, politicians reflect only a segment of a constituency. Once elected, they reflect a broader constituency with far more conflicting interests.

Everyone's First Job is to Keep Their Job

One of the most insightful comments to explain this situation comes from John Mendocha, an author and business consultant. He often says in speeches, "Everyone's first job is to keep their job." That insight helps explain why we have so many problems.

How often have you heard phrases such as "lust for power," or "power is addicting"? There's also another truth to consider. **Even the *perception* of power is addicting!** Politicians don't wield as much power as they think. The People have more power and freedom to use it. We'll go with "power is addicting," to create a foundation for this explanation.

Let's say you're an alert citizen and you see an emerging health problem near fracking wells or oil refineries. You discover that air pollution is causing a high rate of asthma-related deaths in children and elderly persons. You complain to your local government representative or official. However, because you are only one person and one voter, the government official ignores you.

Why Politicians Dismiss Citizens' Solutions

Politicians are very busy, but we rightfully expect a lot from them. They pledged to fulfill certain duties. Never-ending streams of people want something from them so naturally they have to prioritize. That means that a single person will usually not get as much attention as a large group. Large campaign contributors and those with a lot of influence are exceptions.

People with money have influence. However, power is not in the money—but in the People! Elected officials are reactive but don't want to appear as over-reactive alarmists. Politicians might ignore you, even if you're persistent and gather more supporters.

Offering a viable solution to a problem that affects a whole community doesn't guarantee a useful response from any elected officials. They don't like it when citizens have better solutions than the politician has. Politicians want you to see them as leaders, not followers. If you gather more people with you, that may help. However, they are most likely to encourage you to gain support from even more stakeholders. They will wait until you have support of enough people with the greatest amount of influence who collectively represent or affect a *majority* of voters that can, in turn, influence elections. That's how politicians try to keep *their* job.

Such delays are unfortunate for citizens affected by problems and who ask for help from their government. It might take too long to gather sufficient strength in numbers. During that time, the problem continues to get worse. When politicians finally decide it's worth their time to pay attention, the problem is much worse and affects a lot more people. When they eventually respond with a policy or plan of action, the results are often not enough to help people burdened by the problem.

Government Ineffectiveness Explained

The "blue ribbon commission" is a standard feature in our government. It's usually a cross-section of stakeholders invited to evaluate a problem and consider solutions. They're typically an elite group of so-called *experts* who represent various stakeholder groups, but are usually already loyal to the politician.

"Baby steps" are another favorite strategy of politicians. Have you observed in a meeting when someone suggested how to achieve more in less time, less expense, and with greater benefit? Then the leader says, "We have to do baby steps. We need to go slow." You think, "At that rate, our goals are not achievable in my life-time." The world is full of problems and suffering. Who benefits by holding back bigger steps that offer better solutions? Guess what? **Giant leaps are the new baby steps**!

"Best practices" are often not the most affordable and effective solutions. They serve more to fortify a politician's public image and political ambitions. Politicians use less-effective public policies to spread taxpayer money, tax credits, and other benefits to their supporting groups.

That's why you hear that the primary obstacle to solving problems in society is not technological or financial, but political. Politicians feel safer with programs and policies used by their peers. It doesn't matter if their peers' program was any benefit to those affected by a problem, as long as it preserves the status quo. Which policies would you view as status quo?

Governing by "consensus," means working in accord with a majority opinion and often with input from advisors. It's another way to create and continue problems. Decisions made by politicians should be whatever will most benefit people. Tragically, they often don't. The political part obstructs solutions and progress for society because the political futures of decision-makers receive too much importance. Typical politicians use consensus more for *their* advantage than for public benefit. If a better and more cost-effective solution is supported by only a small minority, its use is unlikely. Politicians align with majorities.

Steve Jobs of Apple probably invited input from advisers. Did he wait for a majority vote by them to decide on making his computers, iPods, iPhones, or other electronics? No. He also said he never forced anything. If it was a "square round" (square peg in a round hole) he found another way forward. His vision and problem-solving ability prevailed. He saw a never-ending flow of new solutions. Few people had his vision of possibilities.

Unfortunately, the solutions from such "blue ribbon commissions", "baby steps", "best practices", and "consensus" are usually not the most scientifically proven and cost-effective options. They serve the few at the expense of the many. Often, they're useless.

What's the alternative? Aim high, move fast, be steady and persist! Encourage each person to take the largest steps within their capacity. What was a baby step for Helen of Troy, Steve Jobs, Joan of Arc, Richard Branson, Gandhi, Nikola Tesla, or any other great achiever in history? Did visionary inventor and futurist, Buckminster Fuller, advise people to take baby steps? No. He encouraged them to think big and see what needs doing. He urged them to do as much as they could to solve huge problems and raise humanity's quality of life.

Seize the initiative! Go according to your speed and vision. Don't be held back by someone imposing a slower pace, smaller steps, and less cost-effective solutions just so they can keep *their* job. If you have a viable solution, find a way to move it forward, even if you don't have a majority supporting you!

Justifying Problems for Jobs and Votes

What do all of these tools of governance produce? By the time a typical politician responds to pleas for a solution, and uses such tools above, the problem has already created jobs and business profits. For example, a common promise by candidates for office is that they will increase public safety. They promise to hire more law enforcement personnel, offer more incentives, and upgrade crime-fighting technology. Their big solution is to throw more money at crime in ways that create more jobs or better benefits for law enforcement unions whose vote they are seeking. Denver's jail overcrowding and new justice center were textbook examples.

Politicians do the same for problems with the environment, health, child abuse, and education. Democratic politicians offer programs that benefit unions and other of their supporters. The Republicans offer benefits to corporations and their supporters.

It doesn't matter to politicians how much their policies and programs achieve. It's a way to look like they are doing something about problems. They keep problems going to keep voters working. Again, "everyone's first job is to keep their job."

In other words, if the world seems to be "going to hell in a handbasket", it's because too many people gain too much politically and financially from the *handbasket* industry. That mindset needs to change for us to survive. *RAISE the ZONE* offers a timely mindset *reset* to not only *survive* but *thrive.*

Perpetuating Problems for Job Security

Have you heard the old adage that, "government contractors are paid to look for solutions, not find solutions"? That's also true for politicians and non-profit organizations. In other words, *not* finding solutions ensures funding and job security. Money stops flowing from government coffers when you find a solution. Sales people say, "there are no problems, only opportunities." That means the opportunity to make money from problems.

This perspective obstructs progress. It encourages a lower quality of work. Competition *can* work well between companies. However, an entire industry is not going to work itself out of business by finding better alternatives to fill the same need. The energy and disease care industries are two blatant examples. These mindsets are part of a broader culture that reinforces a problem-based economy. This is another reason why so many problems persist in every area of life.

How Nonprofits Perpetuate Problems

Do you support nonprofit organizations? Are you an active member or administrator of one? Nonprofits deserve great credit for bringing attention to issues that government officials ignore for too long. Many of their achievements have transformed humanity. Non-profits inspired me to find the best solutions for local and global problems.

The risk, is that people expect nonprofits to solve a problem beyond raising awareness about it. Those who protest problems, or offer relief to victims of problems, don't necessarily have the same mindset as those with cost-effective solutions to problems. They have a harder time raising donations if a viable solution threatens to solve their targeted problem. They want to keep their jobs too. I've seen that many times over several decades.

A few years ago, I went to a lecture by a non-profit group from Boulder. It was about coal reserves in America. The speaker knew his topic well. Their research had shown that instead of 200 years' worth of coal left in America, there was only 20 years' worth. He was a big proponent of natural gas as an exit strategy from coal. A coal industry magazine published an article on their work.

The speaker announced that they were potentially receiving a grant from the Rockefeller Foundation. I asked him about clean energy technologies such as "zero-point energy" or "over unity" devices. I inquired how they would fit into Boulder's plan to take over the energy utilities and run their cleaner energy program.

Suddenly, the presenter folded his arms in front of him as his mouth dropped into a snarly frown. He said, "Oh, are you talking about that perpetual motion bullsh*t?" His negative response surprised me. However, they deserved extra credit for meeting with coal miners and asking them if they would like retraining for safer and higher paying jobs outside of the coal industry. The miners affirmed they would welcome that.

How Nonprofits Get Co-opted

I soon found out that one of the Rockefellers had purchased a Russian oil and gas company for $1 billion. It seemed that this nonprofit organization had been "co-opted." The Rockefellers were apparently using this group to help them divest from coal, and profit from natural gas instead, which is still a polluting energy.

Naomi Klein has written about environmental organizations that sell out to corporate interests. Her criticism of corporate globalization and corporate capitalism is well known. Ms. Klein is a Canadian author, filmmaker, and social activist.

Does all of this seem to create conflicts in funding, revenues, and people trying to keep their jobs? Can you understand how both politicians and even nonprofits perpetuate problems? Nonprofit groups should ask themselves frequently if they are doing all they can to end problems that concern them and the public.

Suppression of Clean Energy

Are you aware of ingenious technologies that didn't make it into our lives? Government and private industry have prevented many of them from entering the market over the last 100 years.

Sometimes our government and we *turn a blind eye* to misdeeds of politically influential Americans. Here's one example: William Stamps Farish II was co-founder and president of the American Petroleum Institute, and CEO of Standard Oil of New Jersey. He provided aviation fuel for Hitler's air force, without which, the Nazis wouldn't have lasted as long in World War II?

For "trading with the enemy," Farish paid a fine of a mere $1,000. His company paid a fine of only $5,000, and was required to share patents with the U.S. military. I read about this in a web post from the Massachusetts Institute of Technology in *The Thistle* Volume 13, Number 2: Dec. 2000/Jan., 2001. Many others have publicized these facts too. Standard Oil of New Jersey, later named Exxon, is one of the worst polluters, and among the most politically powerful and profitable corporations in the world.

Whenever someone finds another former guard of a Nazi concentration camp, it makes headlines, regardless of whether the guard was directly engaged in killing Jews. After all these decades, how many Americans know how much American businesses contributed to those atrocities? Farish and his oil company were not the only ones aiding Hitler. Others included IBM, Chase Bank, Henry Ford, Siemens, and more. The U.S. oil industry played a key role in helping Nazi Germany wage war against the world. That's why I'm highlighting its influence.

When Atomic Fracking 'Bombed'

Government has irresponsibly allowed energy development involving extreme risks. Today's natural gas industry will say that fracking has been *safe* for at least 60 years. That timeframe would include using atomic bombs to frack for natural gas. Those atomic explosions were at least twice as big as the atomic bombs used over Japan's Hiroshima and Nagasaki.

That startling disclosure was in an October 8, 2012, Denver Public Library blog article titled *"Project Rulison or: How I Learned to Stop Worrying and Love Natural Gas."* It was about a natural gas company that exploded an atomic bomb underground in Western Colorado to force natural gas to the surface. The Atomic Energy Commission was part of the project and suggested that such gas production could use atomic bombs, "as often as 40 times per day." The project was not profitable because the extracted gas was radioactive and unusable for the public or private sectors.

Do you think those "experts" involved, somehow didn't know that exploding atomic bombs makes things radioactive? Did they know but didn't care? Were workers on site "just taking orders"? Collusion and lack of foresight didn't stop there. Fracking of natural gas makes headlines daily. Images of fracking wells too close to schools and housing developments are disturbing enough. Many politicians and government agencies with friendly ties to fracking companies regularly deny its hazards.

Potentially worse is a little-known program to place small nuclear reactors in neighborhoods throughout the U.S. Of course, "experts" will claim those are safe as well, despite not having a good plan to dispose of radioactive waste. It's the Small Modular Reactors (SMRS) program in the Department of Energy's Office of Nuclear Energy. It's already starting and the public knows almost nothing about this insane plan of local nuclear plants.

For the record, I 'm not opposed to every form of nuclear energy. Breakthrough energy technologies, including nuclear fusion, are going on around the world. However, many researchers and inventors don't trust the U.S. Government and corporations, so they're not bringing their technology to the U.S.

From "Revolving Door" to "Evolving Door"

One outcome of terrorist attacks on the U.S., September 11, 2001, was the USA PATRIOT Act. It regards unlawfully destroying a bridge or mass transit in the U.S. as an act of terrorism. However, if a large corporation destroys or contaminates our water supply, food, or clean energy technologies, the USA PATRIOT Act doesn't include such destruction in its list of terrorist acts.

Exemptions like these happen because special interests place key employees into strategic government jobs. This practice is called the "revolving door" of government. Nonprofit organizations, such as the Center for Responsive Politics and PR Watch, are great resources for in-depth information on the toxic corruption of government and the PR companies it uses.

This toxicity is why citizens need to raise the zone, so we can move out of our problem-based economy. Don't expect traditional career politicians to do it for us because they're not wired for that. Are you?

Imagine if all key employees in strategic government positions embraced the primary goal of ensuring the most benefits for all people. **If more government personnel put our public interests above political gain, the *revolving* door would become more of an *evolving* door!** Each person going through it would increase government effectiveness in serving the public. That will result naturally when we raise the zone because, again, our government is a mirror image of our collective consciousness.

Attempting a Solution-Based Political Campaign

Early in February 2011, I was lying flat most of every day due partly to pain from a car accident in October 2010. It was also due to grieving the loss of my father since December 2010. During that time, Denver's mayoral election was well underway. I had never before thought about running for mayor of any city. However, I knew of solutions to Denver's persistent problems that no other candidate would be offering, so I threw my hat in the ring.

Over several months, about a dozen mayoral candidates were invited to forums. An environmental issues forum was held for a standing-room-only crowd in a large auditorium at the Museum of Science and Nature in Denver's City Park. Environmental groups packed the room. A local Fox TV news reporter moderated the event. He asked all candidates the same questions in rotating order. The next important question was about how each candidate would solve Denver's infamous poor air quality.

The World Health Organization stated that in 2012, about 7 million people died from air pollution. It also said that, "air pollution is now the world's largest single environmental health risk." In 2011, Denver's air quality was so bad that highway signs displayed warnings like "Poor air quality. Go home. Stay indoors."

When it was my turn to go first, the forum moderator asked me how I would deal with air pollution. We each had only one minute to answer. I described a technology that reduced air pollution in Los Angeles within a few weeks. [See more in Chapter 13: "Push Button Air Pollution Solution."] Michael Hancock was the next candidate to answer the same question. He eventually won the election. His response, as I recall, was, "I hate answering questions after Jeff Peckman because he comes up with these answers and I think, oh my God, how am I going to follow that?"

Hancock's comment drew applause and laughs. I appreciated the compliment. It seemed he was open to innovative and cost-effective solutions. His comment started a new trend followed by other candidates. Fox 31 KDVR TV station posted online:

> "At the forum, Tuesday, sponsored by environmental groups and focusing on conservation and sustainability, many of the candidates referred to Peckman's intelligence, some even asking to give their answers to questions before he did so as not to look less thoughtful by comparison when going after."

It was encouraging to see appreciation for solutions I offered. However, scoring points with organizations in the audience was their real goal. Neither Hancock nor any other candidates ever asked me more about the air pollution solution I shared. Even worse, no one from any environmental organizations asked me about my solution then, or since. If *anybody* was interested in reducing air pollution, it would have been these groups.

Forums on education and child abuse were similar. I offered a solution to improve student performance and harmony that I could have made available at no cost to Denver Public Schools. A room full of educators heard it. Again, no one there asked about it afterward. In a forum on child abuse, I mentioned a technology that could reduce stress in homes. It was likely to reduce child abuse. Again, no one was interested.

Becoming mayor wasn't my highest priority. Solutions *were*! However, I did have strong endorsements from popular radio talk show host Peter Boyles; John McConnell, who founded Earth Day; and former U.S. Senator Dean Barkley, who had managed the campaign of Governor Jesse Ventura.

The pattern of resistance to solutions was disturbing. Stakeholders have an interest in an issue or a business. However, they appeared to be less interested in cost-effective solutions, than in politicians who can help keep their funding or profit flowing.

Occupational Hazard of Politics

Politicians aren't proactive leaders seeking to prevent or solve problems in quick and affordable ways. They ignore cost-effective solutions. Instead, they steer taxpayer money toward stakeholder groups. Once the money flows, no politician wants it to stop, even if it shows no measurable results. If they end the money flow, then opponents accuse them of working against children, the elderly, and others needing help. That explains another political adage, "Once on the books, always on the books." The saying, "an ounce of prevention is worth a pound of cure," is useless for politicians.

We the People contribute to declining government by not following that advice often enough ourselves. That's why government budgets and bureaucracy grow, taxes increase, problems continue, and people suffer. Ongoing problems create profitable businesses and industries, i.e., corrections industry, disease care industry, weapons industry, etc. Corporations have a fiduciary responsibility to maximize profits. Pension funds and investment managers also must invest in the most profitable opportunities.

Many critics of government blame corporations for hijacking government policies. There is truth to that claim. Corporations drive government. However, these critics usually fail to mention that many of the corporations they criticize are publicly traded. So, it's a bit ingenuous to blame America's problems exclusively on the owners and top executives of big corporations.

What drives corporations? It's the market – consumers! What drives consumers? It's their needs, beliefs, desires, and so on. What drives consumer needs, beliefs, and desires? It's their mind, body, and emotions. Finally, what drives mind, body, and emotions? It's their consciousness, or rather, the relative coherence and incoherence of their consciousness. Changing that relative degree of coherence will change consumer preferences.

Consumers, and their representatives, rail against corporate influence in government. And yet, consumers also create markets for those corporations and cheer their profits when it benefits consumers' retirement income or investments. Raising the zone, in this arena, means raising coherence to improve consumer choices.

That can be measured as increases in brain-wave coherence on an individual level, and improvements in social indicators on a societal level. These changes in consumers will move markets towards a more solution-based economy. That will then drive corporations to a higher ethic *and* profit, which will then be reflected by government policymakers.

The "Achilles' heel" of democracy, combined with corporate greed and unhealthy choices by consumers creates a perfect storm. How do you untangle this mess? We'll get to that in Chapter 9, "Light Up a Solution-Based Civilization". If you want to raise your quality of life, and have fewer local, national, and global problems, don't wait for government help. Raise *your* zone!

Dump the Duopoly

We must also consider the burden imposed on everyone by the two major national political parties. America's founders predicted great dangers from a two-party system controlling government. Climate change, drug abuse, gun violence, mass shootings, crime, poor health, pandemics, and every other major problem can also be traced to power struggles between Democrats and Republicans in such a "duopoly". Their stranglehold on government holds us hostage. These political parties don't prevent and solve problems.

They gain too much from exploiting problems. They've created a type of Rube Goldberg machine government - an unnecessarily complex and costly matrix that loots and wastes our precious resources but produces minimal results. We *must* free ourselves!

Who is the Most Powerful Person in the World?

How recently have you heard or seen some phrase about LOVE being the greatest power on Earth, or in the whole universe? How many people say that the U.S. President is the most powerful person on Earth? Does that any make sense? Logically, if LOVE is the most powerful force, then the person who generates the most LOVE, not the most hate, would be the most powerful person. If *we* create the most LOVE, then *we* are obviously the most powerful force in the world. That's just simple logic.

For these reasons, there is no value for anyone to panic about who is President of the United States, or head of any other country. That elected leader will still reflect the collective consciousness of its people. What will determine that leader's performance, success, and the quality of our future? We the People will determine these things with everything we say and do! Every day we choose our destiny on an individual and collective level.

Virus of Incoherence

At present, we are a devolved society. Through insufficient amounts of LOVE and coherence, humans throughout the world have created what could be called a "virus of incoherence". This virus is created from incoherence deep within our brainwaves. That's the highest truth about the so-called "deep state".

Chaos and negativity in our collective consciousness result in *vicious* cycles of problems, societal incoherence, and destructive public policy. We need a game plan to inoculate ourselves against this virus of incoherence. Self-government is not a spectator sport. We the People are players, head coach, and team owner.

Transforming Vicious Cycles Into Virtuous Cycles

Coherence and LOVE in our collective consciousness result in *virtuous* cycles of solutions and good public policy. These reinforce each other as well. The best inoculation for the virus of incoherence is whatever increases brain-wave coherence, and therefore more coherent collective consciousness.

That is how we can create solution-based jobs, resources, and profits, to replace the problem-based economy. It should be cheered, and not feared. Another job is easier to find than another planet. It doesn't require all of us holding hands, walking the same path, and speaking the same message.

Do you look in the mirror just to entertain yourself? Or do you look to see ways to improve your image? Problems will continue, even while government gets bigger, until we change ourselves. Who created this mess? We did! In other words:

"The onus of good government is ON US!"

Summary

Digging up roots of our problem-based society has revealed some real gems and a lot of dirt. Does the "Achilles' heel" analogy of democracy help you understand why politicians only follow the majority? How do you view the theory of government as a mirror of our collective consciousness, and the role that coherence plays? There is more digging to do, so hang on to your shovels.

At the risk of seeming overly radical, we need to discuss an unpleasant ongoing situation. The "Great Experiment" of our American democracy is racing to become an extinction event and global crisis. We might need to build a new foundation for a "new model" of government. It's just one possible solution to consider. It's not necessarily the most preferrable solution. However, we'll get to alternative solutions later in the tour.

7

Elevate Your State – Be a Country!

"There is nothing I dread so much as the division of the republic into two great parties..."
– John Adams

Rising Above the Duopoly

Chaos and conflict plague much of the world. The United States of America, in 2026, is on the threshold of an *extinction event* for democracy. The reckless pursuit of power, money, and control, by the two major political parties is driving our society to the edge of a dangerous cliff. We could easily fall into a dark pit of self-destruction, or we could fly into the infinite, radiant expanse of golden opportunities!

Oligarchy and fascism are words often used to describe the current Republican U.S. administration, but there's no laws against those systems. Democrats are often accused of extremism and lawlessness, among other things. The "right" and "left" even accuse each other of terrorism. There *are* laws against *that*!

The United Church of Christ passed a resolution in 2025 stating that it, "... denounces domestic terrorism carried out by ICE agents working without uniforms, wearing masks or refusing to identify themselves."

The Founders of this country warned about such deplorable conditions arising from a dominant two-party system. They also warned of the possibility that a tyrant could be elected as president of the constitutional republic "Great Experiment" that they called the United States of America.

What Would the Founders Do Now?

Both Republicans and Democrats invoke the Constitution of the United States of America. So, why is there so much discord? The following explanation, albeit oversimplified, is meant only to offer a new, or at least uncommon, viewpoint.

Politically, the U.S.A. today is similar to religious conflicts over *orthodox* versus *reformed* doctrines. One well-known example was the bloody "Thirty Year War" between Catholics and Protestants. One group tries to maintain what they view as the purest original doctrine. The other group views founding principles in the original doctrine as guide posts for a changing society and modernized interpretations.

Applied to the U.S.A., it seems the Republican right-wing view is that the U.S. Constitution itself is basically good, except for a variety of Amendments that conflict with their religious and political beliefs and agenda. The Democratic left-wing seems to view the Constitution as an evolving document meant to ensure more rights for the ever-increasing diversity of people, whose various traits, lifestyle choices and origins, fall under the more just and inclusive phrase, "all men are created equal".

It's an ongoing conflict destroying the heart of what the United States of America aimed to achieve. It's highly revered "Great Experiment" has been contaminated by pervasive and deeply penetrating incompetence and corruption from the two major national political parties. Consequently, the entire lab of American democracy is at risk of total destruction. Therefore, it's reasonable to ask, **"What would America's Founders do about this destructive duopoly in our current times?"**

End the Experiment to Save the Lab

Saving democracy's laboratory, that our Founders created, might require ending their 250-year-old experiment and starting fresh. While that might seem self-destructive, or over-reacting, it's what an ordinary research lab would consider in instances of extreme contamination. So, it's worth exploring hypothetically: What would that mean? How could it work? Is there any legal argument that could support ending the "Great Experiment"? If so, what practical solution could achieve that, and still ensure the continuation of democracy and peace among the People?

USA PATRIOT Act of 2001

Customized definitions for "domestic terrorism", as the United Church of Christ created, have little or no legal consequences. In contrast, the *official* definition of domestic terrorism in the USA PATRIOT Act of 2001 has far greater potential legal implications. This historic bipartisan Federal law is the acronym for *Uniting and Strengthening America by Providing Appropriate Tools Required to Intercept and Obstruct Terrorism.* Excerpt:

SEC. 802. DEFINITION OF DOMESTIC TERRORISM.

(a) DOMESTIC TERRORISM DEFINED. — Section 2331 of title 18, United States Code, is amended—

"(5) the term 'domestic terrorism' means activities that—

"(A) involve acts dangerous to human life that are a violation of the criminal laws of the United States or of any State;

"(B) appear to be intended—

"(i) to intimidate or coerce a civilian population;

"(ii) to influence the policy of a government by intimidation or coercion; or

"(iii) to affect the conduct of a government by mass destruction, assassination, or kidnapping; and

"(C) occur primarily within the territorial jurisdiction of the United States."

Who's a Terrorist?

According to the International Center for Not-for-Profit Law:

> "U.S. federal law defines domestic terrorism, but provides no penalties. Instead, the offense of domestic terrorism is state-based, and varies considerably across the country. Thirty-two states and Washington D.C. criminalize the act of domestic terrorism, while the remainder of the states do not."

A recent attempt to re-define the term "domestic terrorist" is the *National Security Presidential Memorandum (NSPM-7)* signed by the U.S. President (POTUS) on September 25, 2025. It was for "establishing a comprehensive strategy to investigate, disrupt, and dismantle all stages of organized political violence and domestic terrorism." The memorandum has been criticized for vagueness and lacking details, while being aimed primarily at opponents of POTUS' unconstitutional policies. Oddly, though, none of the news media, Democrats, or activists made reference to the USA PATRIOT Act in their criticisms, as far as I could tell.

Could America Have a "Terrorist-in-Chief"?

Has any POTUS engaged in "acts dangerous to human life that are a violation of the criminal laws of the United States or of any State". Such violations have been reported.

Do any POTUS' official actions "appear to be intended" to "intimidate or coerce a civilian population"? Threats of, and actual, intimidation and coercion of news media companies, universities, protesters, organizations, or others, have been reported.

Has any POTUS tried to "influence the policy of a government by intimidation or coercion"? The current U.S. administration reportedly attempts to require Federal branches and departments, and state and local governments, to submit to, or align with, the U.S. President's demands. These attempts to influence government policies occur apparently in violation of the U.S. Constitution and some Federal and state laws.

Has any POTUS attempted to "affect the conduct of a government by mass destruction, assassination, or kidnapping;"? Many instances of kidnapping have been reported, including kidnapping of U.S. citizens and others legally in the U.S.A.

It's important to note the word "or" in the above definition of "domestic terrorism". That means, if *any one* of the above criteria of the definition are met by any person, and "occur primarily within the territorial jurisdiction of the United States", then that person fits the official definition of a "domestic terrorist"!

That's not my opinion. It's just a simple matter of comparing any POTUS' actions with the definition of "domestic terrorist" in the USA PATRIOT Act of 2001. If any President of the U.S. clearly meets the official definition of a "domestic terrorist", what does that mean for everyone else?

It means everyone in all three branches of the Federal government is, to one degree or another, engaging in or supporting domestic terrorism. By extension, it also means that any U.S. state or local government providing any kind of funding or support to the Federal government, is also supporting domestic terrorism. Let's take it one more step. It means anyone who resides in the U.S. and pays Federal taxes, or engages with the Federal government in any way, is also supporting domestic terrorism.

America's Epic Predicament

If *any* U.S. President is, as defined by the USA PATRIOT Act, a "domestic terrorist", it creates a new and unique predicament of epic proportions. While that might seem like a shocking claim, it also creates a unique opportunity for a viable and historic solution to America's current constitutional crisis.

So, the reasonable and logical question we should all ask is, **"How can 'We the People' protect ourselves if the U.S. Federal government becomes a domestic terrorist organization?"** It's an urgent and vital question because those who should enforce the PATRIOT Act are part of that POTUS' terrorist organization.

The U.S. Constitution does not address this insane possibility. Domestic terrorism wasn't officially defined by Federal law until 2001 in the USA PATRIOT Act. Fortunately, the PATRIOT Act definition includes violation of a "*state law*". Therefore, to be law-abiding, state governments would have to "dissolve" their bond with the Federal government. States, by default, would also need to assume responsibility for enforcing the USA PATRIOT Act, or their own domestic terrorism laws, to protect "We the People".

As a "government of the people, by the people, for the people," we would need to free ourselves from this potential mess. How many residents of the U.S. grew up wanting to be a terrorist? And yet, we could directly, and indirectly, support domestic terrorism by the U.S. government every day in many ways!

Anticipating Our "Domestic Terrorist" Country

Realizing that the U.S. could someday become a country of state-sponsored, domestic terrorism is vital because it might be:

- Illegal to give any financial or material support to domestic terrorist activities of the U.S. government
- Illegal to harbor or conceal Federal domestic terrorists

Laws regarding domestic terrorism exist in all but 18 U.S. states. Moreover, state laws about domestic terrorism are varied and inconsistent. There are no common definitions and penalties. However, collectively, the various acts that constitute domestic terrorism reveal some potential impacts. For instance, if a U.S. President is a confirmed domestic terrorist, it could possibly be:

- Illegal to pay for any merchandise, services, property use or rental, royalties, or provide any financial or material assistance that would benefit a U.S. President's businesses
- Illegal to provide any financial or material assistance to the U.S. government as an independent contractor
- Illegal for any person or entity to pay Federal taxes
- Illegal to work for the U.S. government

- Illegal for any foreign entities to do any kind of business with the U.S.A.
- Illegal to provide security for a POTUS, his family, cabinet members, or any Federal employees or contractors
- Illegal to allow (harbor) any representatives of the Federal government in any U.S. state
- Illegal to pay tariffs to the U.S., which also constitute "taxation without representation" for American taxpayers
- Illegal to pay any settlement, or even negotiate with such a POTUS in regards to threats and lawsuits

However, enforcing the USA PATRIOT Act means it might be:

- Lawful to seize assets and freeze bank accounts of a U.S. President and his family businesses
- Lawful to freeze any payments to the Federal government from victims of any such POTUS' alleged extortion of universities, state governments, law firms, news media, etc.

In summary, any U.S. President and its government might become subject to all the penalties prescribed by Federal and state laws related to domestic terrorism. If so, such POTUS and his or her administration may effectively give up their authority by engaging in domestic terrorist activities. They would thereby have no legal authority to do anything, or coerce anyone else do anything! It would then be time for true patriots and lovers of democracy to "abandon ship!"

What Could "We the People" Do?

One solution to this predicament would be for a POTUS to not be a "domestic terrorist", as that phrase is clearly defined in the USA PATRIOT Act. That would be better for everyone, including the U.S. President, all employees of the Federal government, all residents of the U.S.A., and the entire world. If that doesn't work, the next strategy might be necessary.

Consider what America's Founders did when they began their "Great Experiment". We could build on their radical idea, which is stated four ways in the first of their two most important documents – The *Declaration of Independence.* Following, are some excerpts – in less than 100 words: [Emphasis added]

1. "When in the Course of human events, **it becomes necessary for one people to dissolve the political bands which have connected them with another**..."
2. "...That whenever any Form of Government becomes destructive of these ends, **it is the Right of the People to alter or to abolish it, and to institute new Government**, ..."
3. "...when a long train of abuses ...reduce them under absolute Despotism, **it is their right, it is their duty, to throw off such Government**, ..."
4. **"...all political connection between them and the State of Great Britain, is and ought to be totally dissolved**..."

Declare Independence Again

The "domestic terrorism" designation would provide more than enough reason for any U.S. state to declare independence from the U.S.A. However, there are plenty of other reasons for such a bold step, even aside from the topic of domestic terrorism.

Our next step needs to be at least as historic as the Founders' Declaration, yet non-violent. However, before getting to that next step, let's walk through a little more historical background. It will help reinforce the logic of gaining independence. The thirteen original colonies were under British rule for 200 years before they had enough of that. We've been under a duopoly for 250 years.

Plato, the ancient Greek philosopher, America's Founders, and modern day thought leaders, anticipated our crisis of today. They also envisioned a solution. I asked ChatGPT about Plato's view of democracy. The response was a scenario that is familiar in the U.S.A. today and probably other nations:

"Overall, Plato's critique of democracy highlights its potential pitfalls, including the risk of manipulation by self-interested individuals and the lack of emphasis on political expertise. He argues for a more structured and knowledgeable approach to governance, where those in power are selected based on their wisdom and ability to govern justly, rather than through popular vote or majority rule."

America's Founders and a Modern Thought Leader

"The alternate domination of one faction over another, sharpened by the spirit of revenge natural to party dissension, which in different ages and countries has perpetrated the most horrid enormities, is itself a frightful despotism. But this leads at length to a more formal and permanent despotism."
– George Washington

"There is nothing I dread so much as the division of the republic into two great parties, each arranged under its leader, and concerting measures in opposition to each other. This, in my humble apprehension, is to be dreaded as the greatest political evil under our constitution." – John Adams

"As the people are the only legitimate fountain of power, and it is from them that the constitutional charter, under which the several branches of government hold their power, is derived; it seems strictly consonant to the republican theory, to recur to the same original authority, not only whenever it may be necessary to enlarge, diminish, or new-model the powers of government; but also whenever any one of the departments may commit encroachments on the chartered authorities of the others."
– James Madison

"You never change things by fighting the existing reality. To change something, build a new model that makes the existing model obsolete." – Buckminster Fuller

Those quotations are from some of the brightest minds in the history of our world. We need to reflect on them and consider how they can guide us in our current crisis. Our ship is sinking.

Going Over the Heads of the Feds

Momentum grows daily for critical solutions to overcome our emerging catastrophe. Do we have an exit strategy? Would it be better to "build a new model" and allow the old one to become obsolete? *Secession* is what comes to mind for most people thinking of "dissolving our bond". That would mean when a state formally separates from the U.S. government to become an independent state.

Both blue and red states have considered the possibility of secession throughout U.S. history. However, either they didn't gain enough traction, or attempted it and failed. One attempt at secession resulted in massive death and destruction. That was the American Civil War. However, the idea is still alive and well. Another inquiry to ChatGPT got the following response:

> "In summary, while a U.S. state could theoretically seek independence, the lack of constitutional provisions, historical precedents, and the need for federal and international recognition make it a complex and unlikely scenario. The legal framework currently does not support the idea of a state becoming a separate country without significant obstacles."

Had ChatGPT existed in 1775, it probably would have stated something similar if asked about colonial America breaking away from King George III. And yet, here we are almost 250 years later with democracy in a constitutional republic because the Founders and colonists did what they knew they had to do. They set their *own* precedent and created a framework. They overcame the obstacles and committed to achieve their goal. That's what we need to do! So, what's the best way to do that?

Imagine if the U.S. government, under a POTUS, violated the USA PATRIOT Act routinely. Again, it could legitimately be called a domestic terrorist organization. Therefore, state governments would need to acknowledge that "**it is their right, it is their duty, to throw off such Government**."

Do states have any practical options for separating from the U.S.A? They need to, because otherwise they would remain as accessories to terrorism by providing financial or material support for, or harboring, such a U.S. President, his or her enablers, employees, and contractors of the Federal government.

Ascension Instead of *Secession*

How can states break away from the U.S. government, given the dismal history of secession attempts? Another strategy would be *ascension* – to elevate states to the formal status of independent countries. That's far better than *seceding* as a mere state. Think of it as maturing eagles leaving their nest. To "build a new model" and "institute a new Government...", we need a fresh flock of fervent feathered founders. That's also a respectful nod to the indigenous leaders who helped America's Founders.

It's likely that international recognition would surge to help U.S. states become countries. Most non-U.S. countries already have an historically low view of our current U.S. administration. They might relish the idea of declaring the U.S. government a terrorist organization because then they wouldn't have to negotiate with the current U.S. President's demands.

"Soft secession" is increasingly thrown around by Democrat leaders and political pundits. That idea is to withhold payments to the Federal government in protest of what they view as overt coercion and extortion by the U.S. administration. So far, such rhetoric has not achieved anything. It's also a very narrow and limited use of government power at the state level. Moreover, both types of secession would keep the People in a perpetual crossfire of the nation-destructing battle between the red and blue parties and states. "Strong ascension" could be the answer!

Both major national political parties have contributed greatly to the current crisis. America's Founders predicted it. It's time that the People rise above these destructive parties and reassert their legitimate power and authority.

250th Birthday Surprise!

Leading up to the 250th anniversary of the adoption of the Declaration of Independence is the perfect time for this next stage of the "Great Experiment". Many of the twenty-seven reasons listed in the Declaration of Independence, for dissolving the 200-year-old relationship with King George III in 1776, also exist in 2026. They overlap a growing list of reasons by the campaign to impeach the current U.S. President again. That includes the biggest trigger of the Revolutionary War – taxation without representation. That's what arbitrary tariffs are today.

However, the biggest potential trigger for seeking states' independence would be if the current POTUS has made the U.S. government into a domestic terrorist organization. None of the Democrats have made that an issue, probably because they want the current POTUS' actions of questionable legality to continue. That way, they can use it against him closer to the next elections.

Is Petitioning the Government Enough?

America's Founders petitioned the government of King George III for various grievances over many years, but to no avail. Their petitions had no legal force and effect then. Today's petitions don't now, at least for petitioning the *Federal* government. And yet, the "right to petition for a redress of grievances", in the First Amendment of the U.S. Constitution, was still brilliant and prophetic. It's the basis for today's *legally binding* petitions which enable the "Right of Initiative and Referendum".

These tools of "direct democracy" empower citizens to institute new ordinances, statutes, and amendments in many cities, counties, and states. That is one of the most effective ways for "We the People" to assert our rightful authority. We have already used that right extensively to achieve voting rights for women, vital child labor laws, environmental protection, reduced taxation, and a great deal more. Those initiatives started at the state level before eventually being adopted nationally.

Are States Competent to Become Countries?

Let's consider the competence and resources of states:

1. The total population of the original thirteen colonies, at the time of the adoption of the Declaration of Independence on July 4, 1776, is estimated to have been approximately 2.5 million people. That included 500,000 slaves.

2. Thirty-six of America's 50 states each have more people than what the thirteen colonies had together on July 4, 1776.

3. California has the 39th largest population in the world among all countries, and the *fourth* largest economy!

4. Wyoming has the smallest population among U.S. states with 587,000 in 2024. Yet, that is larger than each of the sixty-two countries and territories in the world, out of 233, with the smallest populations.

5. The twenty-four states with Democratic governors:

- Contain 57.1% of the U.S. population
- Contribute 63.4% to the total GDP
- Contribute 63.3% of the total Federal taxes collected
- Contribute 78.7% of the total net overpayment to the Federal government among all 50 states. Thus, they subsidize "welfare states" of Republican governors. [Source: USAFacts, updated 11/3/25]

6. States have far more infrastructure, education, commerce, and experience in governance, than the Founders and their colonies had when they separated from King George III.

7. AI and supercomputing can help with a smoother, more rapid, and effective transition for states to elevate themselves to the status of countries. Moreover, transition teams can also solicit input from children with expertise in nation-building video games. That's not a facetious suggestion. The natural innocence, creativity, imagination, and enthusiasm of such children would probably create surprisingly good ideas.

Challenging the Revered Union of 50 U.S. States

America's dire situation today was previously unthinkable to most civilized people. Separating from the Federal government might also seem as unthinkable. Some people might argue that the "sacred union" of the 50 United States must remain intact no matter what. These folks should be asked a question: "Consider if your sibling or child was being severely abused in unthinkable ways by a deranged and violent spouse. Would you urge them to stay in the relationship no matter what, to uphold the so-called "sacred union" of Holy Matrimony?"

The American revolution was either supported or opposed by only a minority of colonists. The majority were "sitting on the fence". The Declaration of Independence was signed by only 56 people, mostly on August 2, 1776, even though it was adopted on July 4, 1776. Only 6% of the population then was allowed to vote because they were white males who owned property.

It has taken most of the last 250 years to get equal voting rights for all U.S. citizens. And yet, that's now at risk of greater obstacles, despite that the U.S. President has no Constitutional authority to regulate elections.

Manifesting a Higher Manifest Destiny

The accumulation of 50 states was not divinely ordained. States added during the first 184 years of U.S. history were largely due to the Manifest Destiny and the Monroe Doctrine. Tragically, the fulfillment of that so-called *destiny* involved a lot of genocide, broken treaties, fraud, racism, oppression, lying, brute force, cheating, greed, and corruption.

Many of those same policies and crimes against Indigenous people and citizens continue to this day. There has never really been "liberty and justice for all" in the U.S.A. We must do better! Our task now is to manifest a higher destiny that offers unifying and practical solutions for healing and prosperity of our nation. By ascending, we can be unified in "liberty and justice for all"!

Moving Upward and Onward

"We the People" are between the proverbial "rock and a hard place". So, the best direction to go is up! Our need now is not to use force, threats, or manipulation for *horizontal expansion* of the U.S.A. into Canada, Greenland, Panama and elsewhere. It is for a *vertical elevation* of competence in self-governance. It's for existing states to ascend to country status. That will open endless opportunities for greater freedom, innovation, self-determination, education, and development of a solution-based society.

These new countries would be freer from the incompetence and corruption of both major political parties at the national level. They could remain independent, or join something similar to the European Union, such as a North American Union with Canada, Mexico, and other North American countries.

These new independent countries could combine the best of their existing state constitutions with the best ideas from the U.S. Constitution. They could also add other beneficial principles that reflect advanced knowledge and greater experience.

Benefits from Ascending to Country Status

When states ascend to the status of countries, they will create whatever government, laws, institutions, and policies they choose. Their choices will align more closely with the will of the People in each new country. America's population is so diverse, that no president can properly represent the best interests and will of the People in all 50 states anyway.

That's obvious when a president, and candidates for president, disrespect entire states, simply when most people of such states hold different political views and interests. Both major U.S. political parties have become too beholden to special interests, often the same ones. It's made both of them almost useless and even dangerous for the well-being of the People. So, let's think about some benefits of states rising to become countries.

For example, the People of each new country could legislate:

- Revenue (taxes, fees, fines, royalties, customs duties, etc.) that would have previously gone to the U.S. government;
- Tax credits, grants, subsidies, and entitlements, funded by taxpayers;
- Elections laws, political campaign contributions, and term limits on elected officials;
- Establishing an Electoral College or not;
- Lobbying and revolving door politics;
- Birthright citizenship, immigration, student/worker visas;
- The right to due process;
- Tariffs and trade agreements;
- Research funding;
- Memberships in the United Nations and other international organizations;
- Patents and trademarks (ending the longtime suppression by both major U.S. political parties of more affordable and cost-effective solutions);
- Consumer protection;
- Emergency funding for recovery from natural disasters; and
- Crime and terrorism, including coercion and extortion by government officials.

That's just a starter list. An abundance of ideas is certain to arise. What great policies are on *your* wish list?

Everybody Wins

States that choose not to ascend will still benefit. This option would free oppressed citizens, and even fulfill the dreams of the U.S. President and all those loyal to him or her. For example, they could end the U.S. Constitution and rule the remaining, albeit much smaller, United States of America for the rest of their life.

The country name could change for the reduced number of states. The new ruler could declare their self to be also King or Queen, Emperor, Pope, and any other honorific titles that he or she might covet. They could even award themselves a peace prize or any other prize they crave. In the case of the current U.S. President, he wouldn't have so much health-damaging stress from his inner rage and from resistance to his policies. He would probably gain better health and live a lot longer.

Raise the Body Politic

Imagine if Colorado, California, Oregon, Washington, New York, and other pioneering states committed to, "...dissolve political bands...", "...institute new Government...", "...throw off such Government, and to provide new Guards for their future security..."! That could be attempted by Executive Order of state governors, or by state legislators or statewide ballot measures.

The phrase "flood the zone" is about lies and propaganda. It's time to *raise the zone* of the body politic with more truth and higher coherence in self-governance. These are vital for building a new model of government. As Thomas Jefferson stated:

> "I am not an advocate for frequent changes in laws and constitutions, but laws and institutions must go hand in hand with the progress of the human mind. As that becomes more developed, more enlightened, as new discoveries are made, new truths discovered and manners and opinions change, with the change of circumstances, institutions must advance also to keep pace with the times. We might as well require a man to wear still the coat which fitted him when a boy as a civilized society to remain ever under the regimen of their barbarous ancestors."

Well, circumstances have changed. "Barbarous ancestors" of future generations are, at present, running and ruining the Great Experiment of America. States can do much better with existing tools, resources, experience, and wisdom.

Too much money and power are concentrated in politicians at every level. That creates easy targets for special interests – courtesy of Citizens United. It's as if both the Republicans and Democrats are battling for the best seats in an elegant dining hall on the Titanic, after it already started to sink.

It's time to elevate states to the status of countries. It's time to move democracy and the well-being of the People out of reach of D.C. politics. Our current crisis is unlikely to be solved by:

- Little-known parliamentary procedures or loopholes;
- Protests that have no legal force and effect;
- Lawsuits;
- Speeches and rallies;
- Democrats controlling all three branches of government; or
- Ridicule, mockery, body-shaming, and name-calling

All of these actions have their respective effects. However, they don't represent the best of our highest nature and destiny. So, what's our next step? We *all* need to ask that big question: **"What would the Founders of the U.S.A. do today?"**

Their Declaration of Independence has vital clues, such as: "Independence", "...dissolve the political bands...", "...abolish it...", "...institute a new Government...", "...throw off such Government...", and "...Absolved from all Allegiance ..."!

Summary

A Constitutional crisis is crushing America. Our Founders' "Great Experiment" has become too contaminated. It now risks destroying the entire laboratory of American democracy. Angry, terrified, and fed-up citizens are pleading for solutions.

Dissolving the bond with the U.S. government, by elevating states to the status of countries, would still uphold the vision of America's Founders. We just need to use our skills and resources more strategically for better self-governance. Courage, vision, and especially *commitment*, are our most important resources!

Our *tour* has taken an interesting *detour*. Perhaps we have stumbled onto an unexpected treasure in our *ascent* to a higher peak of self-governance. If that's where we still want to go, commitment is a magic key, as professed by the Mt. Everest explorer, William H. Murray and quoted in Chapter 1.

We will go another level deeper into how government ignores or buries solutions. The ways you could be contributing to that might surprise you. By understanding these ways, we can unearth the vast treasure of wealth for all of Earth's humanity to enjoy. Through your example, government leaders can quickly become more enlightened and learn to be much more solution-oriented.

8

Harnessing the Technology Supernova

"One key to discovery is to be open to the unexpected."

Where Did We Put Our Keys?

Being a tour guide has extra benefits. One is special access to the Hall of Secrets. Grab your shovels again. We're going to dig for buried world-changing breakthrough technologies. A supernova of innovation is underway that will raise the zone globally.

Technologies that increase coherence are the most important part of this burst of innovation. Why? Because they improve the performance of all other related technologies. Collectively, they can create the "rising tide that lifts all boats". However, before we look at these emerging coherence technologies, let's consider some of the dark history around other kinds of breakthrough inventions that have actually been suppressed.

Do an Internet search on "suppressed technology." There is more to it than suppression by governments and corporations. You might be contributing to this unknowingly. The *suppression* of beneficial technologies, by and for special interests, has been one reason why people mistrust corporations and our government. The *oppression* of entire races and populations is yet another part of our planetary history and old paradigm.

Have you attempted to present a solution to elected officials and been ignored, or even obstructed? I'll again share some observations and conclusions drawn from my own efforts to introduce solutions to not only government, but to non-profit advocacy groups. Feel free to draw your own conclusions. I welcome alternative points of view if they help empower citizens even more.

Political Injustice for Criminal Justice

On the heels of the *Safety through Peace* ballot initiative, I began to promote Farwell Brain Fingerprinting, "a scientific method to discover the truth, identify criminals and terrorists, and clear the innocent with extremely high accuracy by measuring brainwaves."

TIME magazine had honored inventor, Dr. Larry Farwell, in *TIME 100 Top innovators of the Century,* "the Picassos or Einsteins of the 21st Century." His technology had been tested by the FBI, U.S. Navy, and CIA. Dr. Drew Richardson, formerly the Director of the FBI's counterterrorism office, was so impressed that he quit his FBI job after 22 years of service and joined Farwell's team.

Thomas Frey, Executive Director of the DaVinci Institute, and I created a practical strategy to promote Farwell's technology. We recruited about a dozen leaders from law enforcement and the Colorado State Legislature. We called our startup group the Colorado Brain Fingerprinting Task Force. That group included Chiefs of Police from Aurora and Longmont, and the Director of the Police and Fire Academy for Aurora.

Two other members were Colorado State Senate Majority Leader Ken Gordon (D) and Colorado House Minority Leader Joe Stengel (R). Both of them jumped onto the task force within just a few minutes of my introducing the Brain Fingerprinting technology to them. The invention was effective in improving criminal justice, so it appealed to lawyer, Sen. Ken Gordon. It could reduce the costs of criminal justice, so it appealed to fiscal conservative, Rep. Joe Stengel.

The Douglas County, CO Office of Economic Development expressed strong interest in having the national headquarters of Brain Fingerprinting Labs in Douglas County. It was expected to potentially bring 300 jobs and cutting-edge forensic technology to Colorado's criminal justice system. All was going full speed ahead. Then it came to a disappointing end.

Political Challenges to Truth Technology

Rep. Stengel introduced a bill in the Colorado State Legislature to support Brain Fingerprinting as a way to reduce expenses for criminal justice and incarceration of criminals. Sen. Gordan was ready to support it in the Colorado Senate. Ironically, the polygraph (aka lie detector) industry, had been spreading false information to discredit this competing technology. Brain Fingerprinting had already met the Daubert Standard and had been admitted as evidence in court. It has been instrumental in exonerating a wrongfully imprisoned person, and identifying a serial killer that had previously eluded prosecution.

When the Democrat-controlled committee presented the bill by Rep. Stengel, their review cited misleading information from Farwell's competitor. They didn't allow any comments outside of committee members. It was a quick presentation. They voted against the bill. When I immediately shared that news with Senator Gordon, who was waiting in his office, he replied, "they just wanted to stick it to Joe". Partisan politics killed a great solution.

Greenwashing in Politics

Environmental issues also have politicians who passionately allege to support clean air quality. However, that's only to the extent that it benefits their electability and campaign supporters. Nine of Colorado's Front Range counties have violated EPA air quality standards for the last 17 years. This has all been during the dominance of Democrats in the governor's office and state legislature.

In the spring of 2024, I informed the Colorado Energy Office about two technologies that were more cost-effective and affordable for improving air quality, than anything else being used in Colorado. I shared the same information with Rocky Mountain Institute (RMI), known as a champion of environmental issues.

Neither the Colorado Energy Office nor RMI took interest in the technologies I shared. They merely referred me to various competitive mentoring and startup company pitching opportunities. RMI and Governor Polis are intent on electrifying everything possible, to move away from fossil fuels. A flaw in that strategy is that it completely ignores *electropollution* and its role in causing major problems for health and society.

One of the technologies I introduced could reduce emissions of gas and diesel vehicles enough to give them lower life-cycle carbon footprints than electric vehicles. This same technology can also convert methane from natural gas, landfills, waste treatment plants and agricultural waste, into hydrogen and graphene. It can even convert CO_2 captured from the air, breweries, etc., to high-value graphene. Research quality graphene is many times more valuable than gold! Why spend billions of dollars to pipe and bury vast CO_2 treasures in the ground?

The new "wonder material", graphene, is being researched by 35,000 people around the world. It's expected to improve the performance and sustainability of, for example: solar cells, wind turbine blades, concrete, asphalt, microchips, vehicles, and aircraft. It has also shown great promise for purification of sea water, chemically toxic water, and even radioactive water. Another important benefit is that it can improve batteries – possibly even replacing hazardous and fire-prone lithium.

According to a newswire by graphene producer, Avadain, "America's No. 1 tech futurist and legendary stock picker George Gilder predicts graphene will have a $11 trillion macroeconomic impact." Bank of America and Goldman Sachs have predicted that hydrogen will have $11 trillion in global economic impact by 2050. Such projections can change, but must be considered.

These technologies are a natural and obvious fit for energy and environmental policies. That includes for the highly touted Colorado Greenhouse Gas Pollution Reduction Roadmap 2.0 and related Colorado legislation. Unfortunately, what struct me about the Greenhouse Roadmap is that it read more like a "White House Roadmap" for someone's campaign for president.

The GHG Roadmap did not mention graphene a single time! However, it did mention Rocky Mountain Institute, or RMI, thirty-eight times. It also mentioned hydrogen four times and "green" hydrogen one time. That was in reference to Fortescue Industries, an Australian company that makes green hydrogen through electrolysis, which uses five to seven times more electricity than other ways to make hydrogen. It seemed that RMI was largely the author of Roadmap 2.0. Fortescue is one of the strategic partners of RMI. I suspected that the Colorado GHG Roadmap was highly political. My suspicion was confirmed by someone in Colorado government.

To be fair and balanced, I also introduced these technologies to the Colorado Oil and Gas Association, which advocates for the oil and gas industry. At first, it seemed logical that they would welcome the information. The technologies I suggested would increase profits for the oil and gas industry, while improving air quality more than would happen otherwise. However, they were also not interested. That surprised me. Then I realized that their advocacy operations depended on funding from the oil and gas industry to fight what they regard as *hostile legislation*.

Endorsing the new technologies that I informed them of would reduce the amount of hostile legislation in Colorado. That would mean the Colorado Oil and Gas Association would have a harder time extracting money from the oil and gas companies to support their public relations and litigation activities.

You'd think that Colorado politicians and non-profits would get excited about advances in technologies that are emerging leaders of potentially multi-trillion-dollar markets. Instead, the above examples illustrate how politicians, and even non-profit organizations on both sides of any issue, put their own interests above the needs of the People. They like to present themselves as champions of some problem of concern to the People, but then actually perpetuate the problem instead. They like to advocate for "sustainability", primarily to sustain their political careers and fund their non-profits. That makes real sustainability for the People, and environmental quality, a lower priority for them.

However, it's better to presume that the decision-makers for these politicians and non-profit organizations simply mean well and are doing the best they can. After all, we now already know that government is a mirror image of the collective consciousness of the People.

Nevertheless, it's useful to discuss this current reality in order to understand why it's necessary to raise the collective consciousness of the People. That's how government, non-profits, and the private sector *boats* will be *lifted* even more by that *rising tide*, to benefit our society much more.

Does Suppressing Technology Benefit the People?

Public benefits and suppressed technology are not conflicting ideas. We're moving out of that into a new paradigm with a new mindset. "*Compression*" can result from either suppression or even oppression. It's part of the ebb and flow of life and society—the waves of development of humanity. When you press your hand against water in a pool or pond, etc., it creates a wave with a peak and a trough. That wave can move across water in simultaneous states of compression and expansion.

Have you ever suppressed something in your life? Did you hold back out of fear, or doubts about your ability to succeed? Think of standing on a diving board before you bounce to create momentum. Then you bounce and compress the board to spring through the air and into the water. Have you ever pressed yourself to achieve something – maybe an athletic feat or some higher level of academic or career success?

Releasing that pressure brings expansion of your intelligence, energy, and creativity to achieve your goal. It takes you to a higher level. From there you could start another wave, or series of waves, expanding even more. That is the nature of life—to grow in LOVE, happiness, health, success and so on. Therefore, on a societal scale, our collective evolution will see periods of great suppression, as well as oppression and compression, and then great leaps of fresh expansion in every area.

This burst is parallel to how a black hole devours and compresses matter and even light. It takes kinetic energy of a star and compresses it into a small area of potential energy. At some point, compression gets so extreme that gamma rays burst out of the black hole to spread out and seed life elsewhere, according to some astrophysicists. Potential energy in the compressed state again becomes kinetic energy.

This analogy offers a vision of how humanity could soon burst out of its suppressed or oppressed state. Then it will seed new technologies and solutions on an epic scale in record time. It could also inspire new ideas and policies that support humanity's great transformation. We need intention, desire, and will to realize that grand opportunity for a *Luminous Age* of humanity.

This chapter might create anger and frustration, which can be useful, but only if it's properly channeled to fuel positive action toward useful solutions. Anger and frustration are not solutions in themselves. Noble intentions must follow them which, in this case, is to achieve a healthier and more effective government.

We the People are the body politic. We desire great health and function of our bodies. So, it's in our best interest to want our government, that we established, to organize great opportunities for all of us.

A vital and healthy body can achieve much more than a weak and sickly body. Just as a body fighting within itself creates horrible suffering and disease, a country's population fighting within itself causes widespread societal problems. These, in turn, can negatively affect our personal lives and the whole world.

Time for the Burst of Technology

Suppression of breakthrough technologies in energy and health care must end to ensure our very survival. Devastating wars, pollution, and disease, must decrease too. Many of these problems relate to protecting our production and use of oil, gas, coal, and nuclear energy industries and their respective investors.

The motives of government and corporations to suppress beneficial technology have a foundation in public support. Does that statement surprise you? Many investors seek the most profitable investments, regardless of their impacts on humanity. Competition is undermined with the help of political favors. Politicians are obliged to fulfill their campaign promises of creating and protecting jobs, even if they are in problem-based industries.

Many breakthrough technologies offer better opportunities to achieve job-creation objectives. However, entrenched entities don't have enough vision to see that. Therefore, investors of all types, directly and indirectly, deny themselves and the rest of us full benefits of extraordinary new technologies.

Admittedly, many mechanisms are in place to help ensure public safety. Some inventions have been potentially hazardous. Some might not have ever worked as claimed. Misuse of inventions is another possibility. Using advanced technology requires higher wisdom and expanded thinking to consider full impacts of new technologies. Such wisdom is often lacking.

An Alternative History and Future

Polluting and hazardous industries probably would not exist today without suppressing breakthrough technologies. However, they do exist and thrive. Their life support results from artificial manipulation of America's highly touted "free market." An unhealthy alliance between government, industry and a handful of wealthy individuals contributes to that condition. We must transform this destructive alliance into a healthier one for humanity.

Suppressing advanced beneficial technology perpetuates and ensures a problem-based economy. Nikola Tesla was just one of many inventors whose technologies were suppressed, or in some other way, undermined. However, suppression of *his* inventions probably did the most damage to humanity.

Tesla invented a world-changing alternating current induction motor, robotics, wireless communications, radio and remote control. He envisioned the Internet, smart phones, and GPS. The plots against Tesla by J. P. Morgan and Thomas Edison were deplorable acts of greed. Their consequences can be seen now in catastrophic global air pollution and climate change.

Which other ways have suppressed technology? They include obstruction of patents and unlawful destruction, confiscation, or hiding of technology. Would you believe they also result from self-defeating activities by inventors themselves? We'll get to that too.

How "National Security" Abuses Obstruct Innovation

Do you know how the U.S. Patent and Trademark Office (USPTO) buried thousands of technologies during application for patents? Claiming national security interests, and citing a Secrecy Order under Title 35, U. S. Code (1952), sections 181-188, is one tactic. This law allows withholding patents by requiring secrecy for specific inventions. It's reasonable to monitor new inventions that could potentially threaten national security. However, that law has been abused.

In April 2013, *Wired Magazine* addressed this topic in an on-line article titled, *"Government Secrecy Orders on Patents Have Stifled More Than 5,000 Inventions."* It described how criteria used to require an inventor to keep his invention secret were vague and shrouded in secrecy. Secrecy orders can apply even to inventions that could be viewed as detrimental to national *economic* security of the United States. Under that umbrella, a *secrecy order* could stifle any energy technology that is cleaner and less costly than fossil fuels, because of their impact on the national economy.

Is the Patent Office Patently Unfair?

The USPTO also used an *illegal* policy to delay patent review for years without explanation. Their secret "Sensitive Application Warning System" (SAWS) was active for over twenty years before being discovered and publicized by a Denver patent attorney. Its discovery resulted from an informal comment by a USPTO patent examiner who mentioned a "secret program" at the USPTO. The criteria were vague, broad, silly, and even absurd for unjustifiably delaying selected patent applications.

The number of applications targeted, roughly equaled patents held by the Wright Brothers, Nikola Tesla, Thomas Edison, and Steve Jobs. Following is a partial list of targeted applications taken directly from internal USPTO memos:

- "Anti-Global Warming devices or any other device operating at the global scale"
- "Motor, Power plant, or other device which is self-sustaining (perpetual motion) or appears to violate the laws of chemistry or physics
- "Claiming prevention or curing of diseases which were previously considered impossible to prevent or cure."
- "Applications which would potentially generate unwanted media coverage (i.e., news, blogs, forums)"

How secretive was it? Neither the USPTO's website, nor its highly detailed 1,500-page *Manual of Patenting Examination Procedure,* mentioned the SAWS policy. The patent examiner said he was not supposed to talk about it. The law firm used the Freedom of Information Act to request details about SAWS in October 2014. The USPTO denied two of their requests before giving in. The law firm's exposure of SAWS appeared in mainstream news:

> ***"The U.S. Government Has a Secret System for Stalling Patents,"*** by Alyssa Bereznak, National Correspondent Technology on December 3, 2014. – Yahoo! News.

The USPTO tried to justify its SAWS program in January 2015, and claimed to have ended it in March 2015. Keep in mind that the USPTO's illegal and secret policy was underway during *both* the Democratic and Republican administrations!

Protecting Extraordinary Technology

According to the USA PATRIOT Act, intentional destruction of an energy facility or a component of mass transportation is an act of terrorism. Don't breakthrough clean energy and transportation technologies deserve as much protection? After publishing the first version of this book in October 2017, I drafted a ballot initiative for "Protecting Extraordinary Technology". [See a summary below and the full text on page 216] The question that could appear on an election ballot is called a "ballot title":

> "Shall the voters for the City and County of Denver adopt an Initiated Ordinance, and in connection therewith, require the city to issue patent protection for extraordinary technologies, and affirm the legality of research, development, manufacturing, marketing, sale, possession, and use of extraordinary technologies?"
>
> Yes___ No___

The proposed binding provisions were as follows:

> **Sec. ___. Protecting Extraordinary Technology**
>
> 1. The City and County of Denver shall provide expedited issuance of patents for extraordinary technology, giving highest priority, without being a prerequisite, to categories of patent applications that were, or continue to be, unfairly or illegally obstructed or delayed by the U.S. Patent and Trademark Office. Funding for processing patent applications and issuing patents shall be obtained through application fees, or otherwise, as determined by the City and County of Denver.

2. The City and County of Denver shall affirm, through a public proclamation and general policy, the legality of research, development, manufacturing, marketing, sale, possession, and use of extraordinary technology, regardless of patenting status.

You might be aware that the U.S. Patent and Trademark Office had a satellite office in Denver, which closed permanently in 2024. It's also true that the Federal government preempts lower levels of government in patent issues. However, that preemption applies only to patents issued at the *Federal* level.

A city or state-level patent office would keep national politics and bureaucracy out of the way of new technological innovations. The proposed Denver ordinance would mean that enforcement of patents and protection of their inventors, along with the use of breakthrough technologies, would be restricted, for example, to Denver, the state of Colorado, or California, etc.

My draft initiative for protecting extraordinary technologies was formally registered in the State of Colorado, Ventura County, CA, and Denver during 2018-2019. Not surprisingly, even though the governing politicians were aware of it, none of them in any location asked me about it. The "Power of the People" seems to terrify them and they don't like being upstaged by voting citizens.

Inventors' Self-Suppression

Brilliant, courageous, and honest people are making crucial breakthroughs in technologies around our world. Sometimes, however, inventors suppress their *own* technologies, instead of governments or corporations suppressing them. Inventors can become fearful of getting undermined, harmed, or worse. It can cause them to lose time, money, and their commitment.

Are you suppressing your genius out of fear, doubts or another negative mindset or behavior? Suppression of anything can become a springboard to the next stage of evolution for our civilization. A springboard is inside every person. Do you feel something within you that has potential energy and is ready to burst forth? Do you feel power and desire building up and accelerating?

To get that going, remember the power of intention and commitment. It starts as a first step. Expanded minds can create new technology, but they can also repurpose existing technology. Adding in wisdom of the heart guides our use of technology along a path of higher benefits for everyone.

Words of Caution for Inventors and Their Supporters

Inventing is not immune from problems common in other businesses and human behavior. My observation is that inventors' disdain for the fossil fuels industry energizes passion for some people to create a new energy breakthrough. And yet they find common sense business practices to be challenging.

The negative view and mistrust that some of them have of others can become an obstacle to their success. Some inventors have misrepresented their achievements to attract investors. Some can't adopt a business-friendly mindset when they get close to releasing their product into the market. Internal conflicts within invention research and business groups have often been huge obstacles for vital new technologies getting to market.

What can solve these issues? One solution might be for them to check their moral compass frequently. It would also help if they stepped back from their notion that *their* invention alone will solve all related problems. That attitude is somewhat typical from what I have experienced in my interactions with many inventors. Their unrealistic level of confidence might keep them going through hard times, but it can ultimately be self-defeating.

Inventors need encouragement to confront daily business challenges. If you want to invest time, money, or expertise in a breakthrough technology, then find out some facts first. Find out about the people involved, as much as vetting the invention itself. Do basis research about competitors in the same market. Read news about developments and successes in the same industry.

How is *Your* "Vision"?

Another source of suppressed or unrealized breakthroughs in technology is a cultural mindset. People with strong credentials and influence create this mindset by ridiculing new inventions and ideas. They lack vision to see expanded possibilities. How does your vision compare with doubters and debunkers?

Have you ever seen a collection of quotations from famous or respected *smart* people who have said silly things about new technology? Some scientists claimed that inventions such as airplanes, radio, electric lights, talking pictures, telephones and computers had no future use or market. Sometimes even the *inventors* of these technologies dismissed their future.

According to some "experts," Albert Einstein and Steve Jobs had too little intelligence or education to succeed. What do you remember most? The geniuses who disproved doubters, or do you remember the doubters and skeptics more? Amazing inventions, and people that made them succeed, are the most memorable, are they not? Limited thinking is neither inspiring nor energizing. It's not even worth remembering.

Growing a "Y.E.S." Culture of Solutions

Ideas need a fertile ground of support and belief. Steve Jobs said it was important grow a "Y.E.S." culture of solutions where "You Expect Success!" Is your "yes" powerful enough to defeat your "no"? Steve Jobs had doubts as well, despite his great creativity. Even Jobs experienced failures and those "dark nights of the soul," only to reinvent himself and Apple Computers. His visionary technologies reinvented human progress.

Have you found yourself bound by viewpoints and paths that no longer suited a future you wanted to create? Did you hold on to them merely to preserve a past habit? Was it hard to let go and approach it with a fresh perspective? That's natural. We invest in an idea or belief that's an expression of ourselves.

When we over-value a concept, we have less resources with which to imagine something else, even if it's better. We get so identified with the lesser concept that letting go almost seems like death to part of who we are, and our existence can feel threatened.

Is it any wonder that change comes slowly to our society? If you start with a "no," nothing useful will happen. Change has already been defeated and disproven without even seeking its value. For those who grow out of limited and single-option thinking, abundant freedom and opportunities await!

All right, here's a little tour guide recreational moment. If you ever leaped to more expanded thinking and reinvented your view of possibilities, how did you do it? How did you feel *after* doing it? Did you feel even more like your true self when you made your breakthrough? If you did, then maybe you can relate to another of my father's quirky sayings. He said, "I feel a lot more like I do today than I did yesterday." Ponder that for a moment.

Try the "Y.E.S." experiment. Again, that means "You Expect Success." Hold that expectation in your mind. Now think of a problem that you have thought was impossible to solve. Challenge your limited thinking. Be creative. Be flexible. Be innovative. Let your mind flow freely. Are any solutions bubbling up? If not, don't worry. Come back to it later. Find a way to relax.

Archimedes' had his legendary "Eureka" discovery of water displacement while relaxing in a tub of water. Unexpected breakthroughs by others resulted in Velcro, penicillin, Post-It notes, microwave ovens, graphene, and much more. One key to discovery is to be open to the unexpected. Your essence is truly an infinite wellspring of creative intelligence.

Notice if your intention to solve a problem starts giving you clues of where to go next. It's just an experiment and the answer depends on how you ask the question. You already succeed, merely by trying it. Robert Oppenheimer, "father of the atomic bomb", once said, "sometimes I ask a question and I let the math answer it. Sometimes I asked a question and I answered it, and I let the math prove it."

Embracing Visionary Ideas through LOVE

Limited thinkers ignored, misunderstood, ridiculed, tortured or killed great visionaries and champions of humanity throughout history because they threatened the status quo. We can think of Moses, Krishna, Buddha, Mohammed, Jesus, Joan of Arc, Galileo, Nikola Tesla, and others.

These extraordinary people seemed different. They challenged limited thinking and views of their time. It might have taken many decades, centuries, or even millennia, for humanity to grasp the extraordinary contributions of these visionaries. Throughout that time, society denied itself the great benefits of their achievements.

That kind of limited vision has been counterproductive to us as individuals and to our entire civilization. We don't need to allow it to continue. LOVE will open us to vast opportunities. By LOVE, I mean unlimited LOVE, the greatest force in the universe. Albert Einstein expressed his view and conviction that LOVE creates and governs everything in the cosmos.

This discussion is not about personal romantic love. Without such universal LOVE, which is inclusive, we'll reject new ideas out of fear of something different and a sense of separation. We can choose to be more open-minded. LOVE and coherence are the keys. We'll explore these keys more in our tour at Chapter 10: Truth, Coherence, and LOVE – The Antidotes to Stress.

Politicians Are the Last to Know

In 2015, I sent emails to politicians at local, state, and federal levels about ongoing suppression of clean energy technologies. The only politician who responded was a progressive Colorado state representative. He actively opposed fracking and supported clean energy. He replied by email, "I know the fossil fuel industry has benefitted from a century of government subsidies and special treatment; however, I'm not clear how alternative fuel research has been actively suppressed or disrupted." No surprise there.

His response shocked me, even though I give public servants some benefit of the doubt. Again, we should assume they are doing the best job they can. Our thoughts and words have power. It's a waste of time to become obsessed with distrust or fear of our government. That disrupts harmony in the relationship between citizens and government. If we hate politicians, that will decrease their effectiveness. Negative attitudes are part of an old paradigm. It doesn't serve our purpose as we move into a new era of better governance. In that transition, better citizenry is required too.

There are always dedicated public servants with genius and a higher ethic. They raise the bar and elevate governance by their example. They rarely get publicity and recognition they deserve. News media focus too much on drama, scandal, and conflict.

That being said, we still need to take responsibility for getting things done ourselves when politicians seem hopelessly lacking information, or are incapable or unwilling to do what needs to be done. We're part of the rising tide that lifts all boats. We're all connected. We all need to expand our hearts and minds.

Genius is not only about brain power; it's about heart power too. The more we do to lift ourselves up, the more we lift up others. That includes politicians. **Research has shown that respecting someone brings us measurable mental and physical benefits.** It does so for the other person as well.

Summary

Both Democratic and Republican administrations of the U.S. government have secretly and illegally suppressed breakthrough inventions that could have greatly reduced damage from pollution, climate change, and disease. Their primary strategies were the U.S. Patent Office's Sensitive Application Warning System (SAWS), and the "Secrecy Order", under Title 35, U. S. Code (1952), sections 181-188. Obstacles to bringing breakthrough technologies to market have also been caused by industrial sabotage, non-profit organizations, and even inventors of extraordinary technologies.

All of this suppression and these obstacles still reflect the collective consciousness of humanity. Technologies which "raise the zone" by increasing coherence in collective consciousness will therefore be vital for harnessing the technology supernova. We must end this suppression and be open to new possibilities that serve the highest aspirations of humankind. Those are our tasks to achieve, as ordinary citizens, for ourselves and our world. Okay, now let's tour some philosophical elements and practical impacts of our mission to *raise the zone.*

9

Light Up a Solution-Based Civilization

"Educating the mind without educating the heart is no education at all."
—Aristotle

Are you ready to move from the Hall of Secrets to the Garden of Opportunities? It's amazing what can grow in fertile ground of an open mind with some seeds of imagination and careful nurturing. You can plant a new civilization with readily available seeds of knowledge. Are you into meliorism, the belief that humans can improve our world? Our planet is full of meliorists.

Many signs show a shift towards a solution-based civilization and economy. However, it's not enough to replace a problem-based economy with a solution-based economy. The very word *economy* has been limited in definition to be mostly about jobs, stock markets, investments, economic engines, trade and money.

Life on Earth is far more than these limited ideas of economic systems and money, and so is this chapter. Your human resources are your greatest wealth. Your most powerful resource is an ability to experience and express LOVE. Raising everyone's life to the highest level is the real purpose of economics. Transitioning to a solution-based civilization is a big leap. Preparing for that will mean first passing through a cluttered "Attic of Outdated Ideas". The challenge is that many people have a hard time letting go of ideas that are obsolete or are no longer a good use of resources.

Cleaning out this attic will require clarifying and correcting some of the most influential ideas that have kept us in a problem-based society. That will lay a firm foundation for not only a new *mindset* but also a new *heart set*. It demands answers to such eternal questions as "What were we thinking?" Part of our problem was that we were thinking too much and feeling too little.

If we aim to advance humanity and uplift civilization in a sustainable way, LOVE and coherence, coupled with expanded thinking, can accomplish that. Nevertheless, a short review of history will reveal where we misread some signs and slid into a ditch. We'll get into practical ways to develop our solution-based civilization with a more illuminated economic perspective.

Conscious Capitalism

Patricia Aburdene, Ph.D., highlighted positive new economic influences in *Megatrends 2010 – The Rise of Conscious Capitalism.* She masterfully shows evidence of emerging corporate trends toward more socially responsible and ethical business practices.

Executives and board members are recognizing their need to consider both *shareholders* and all *stakeholders* affected by their practices and products. She also highlights the role that more conscious and ethical consumers have on markets and on influencing corporations toward a higher ethic.

Dr. Aburdene followed up with her book, *Conscious Money - Living, Creating, and Investing with Your Values for a Sustainable New Prosperity.* That book is more for consumers and small investors. Her highly praised books show that our world is shifting to a higher consciousness of investing by businesses, institutions, and individual consumers. In all cases, the private sector is leading the way – not the government. She views "enlightened self-interest" as the key to making it all work.

Enlightened Self Interest vs. Selfishness

There's always a debate about what drives our economy. Dr. Aburdene takes issue with those who believe that greed is needed for capitalism to work. She argues that enlightened self-interest is better at creating abundance and prosperity. She sees the old idea of greed fueling capitalism as obsolete.

Does greed seem selfish to you? Does it seem worthwhile that greed, by "fueling capitalism," resulted in building industries and empires that provided masses of jobs and economic engines? To clarify these viewpoints, I'll focus first on advocates of *free market* capitalism. These advocates strongly urge getting government out of the way of business. Their *philosophy* praises deregulation and independent thinking in service so one's happiness, as the primary guide for conducting business and one's life.

To the extent that they advocate personal responsibility, productivity, achieving excellence, and individual rights, they do a great service to individuals and society. Those are all good things. These concepts are not new. They're essential principles upheld in cultures throughout our world for millennia. *RAISE the ZONE* also strongly endorses those principles. However, it doesn't support the incomplete philosophy of unregulated greed.

Flawed Notions in Free Market Philosophy

Any belief can become a tool to uplift society, or a weapon to destroy it, depending on one's intention. You see, there're flaws in free market philosophy. That's especially evident in the concept of using self-interest as the primary reference point, without a broader context. Too many people have the notion that altruism is bad for profit. They use that notion to justify greed and profit at the expense of the quality of human life and the environment.

Do free-market advocates have *any* altruistic perspective? Does altruism enter into their philosophy or life at all? Is it all *black and white?* What's the foundation of this narrow view? How would *raising the zone* relate to this philosophy and our world?

Are Selfishness and Altruism Opposites?

Some people espouse that all would be fine if everyone simply did their best to satisfy their personal and business self-interests, without government interference. They believe that if businesses use self-interest without restriction, the economy will thrive. They endorse generating prosperity according to the effort of each person. Such views have value and have gained a loyal following. However, are those principles good enough to justify every effort in business? If not, where's flaw in this philosophy? It's that many people accept "greed is good" because it's simple, without first understanding its full consequences.

It's one of the mistaken uses of Occam's razor. The "simplest explanation" does not always lead to a deeper understanding. A better approach is to view principles in more expanded contexts. That means socially, spiritually, collectively, and economically. A couple of definitions will be useful here:

- *Selfishness* usually means someone is caring only for his or her self, and often at the expense of others.
- *Altruism* (or selflessness) is a concern for the well-being of others, even at one's own expense.

Selfishness and altruism are opposing behaviors in the view of most people, at least on the surface. In a bigger context, their differences depend on your viewpoint, which is often expressed in debates on the differences between capitalism and communism. The House Un-American Activities Committee hearings, mostly in the 1940s and 1950s, was an example of attempts to identify and purge suspected "communists" from Hollywood.

Among other things, that committee blacklisted Hollywood actors and screenwriters who promoted worker rights in the film industry. Some ambitious politicians labeled Hollywood activists as communists. Ultimately, it was a display of differing philosophies of governance. Now, we easily accept the rights of these formerly exploited workers and their unions.

As a refresher, let's look at self-interest an altruism with reference to a few definitions related to governance:

- *Communism* is a way of sharing resources of individuals and society for the benefit of everyone in the society.
- *Collectivism* is similar to communism in its focus on the interests of group well-being and its various systems for politics, education and the economy.
- *Capitalism* emphasizes private ownership and profit as the foundation of an economic system and the production of goods and services.
- *Individualism* advocates the rights of individuals over the government.

In western society, we typically think of communism as the political means of oppressive and brutal dictators. It has a lot to do with making public ownership of land and property for the good of the whole society. However, religious and scripture-based communal groups, communities, and monasteries have *voluntarily* shared property for the common good on smaller scales. Collectivism and communism have similar aims and seem opposite of capitalism and individualism.

Many people have endured one kind of *forced* "altruism" in the Soviet Union. Those promoting capitalism are not necessarily opposed to kindness and charity. Rather, they view such acts as having at least some self-interest. A different kind of self-interest was also a motivation for communist leaders in Russia and the U.S.S.R. Despite trying to appear as champions of the people, the workers, they put their quest for power first.

However, to often, the *heart* of free market capitalism's narrow philosophy is missing. It puts self-interest and rational thinking above, and almost to the exclusion of, feelings and heart. That kind of view rejects *knowing* from the heart. It gives too much value to objective calculations of short-term profit at the expense of everything else.

Individualism, selfishness, and capitalism are logical to group together. They reinforce each other and share a common root of being in service to one's self-interest. Free market capitalism based on narrowly defined self-interest rejects any sort of collectivism, altruism, and communism. These are logical to group together because they share a common root of being in service to others.

At a more superficial level of perception, we're separate from each other. Therefore, it's not surprising that "you vs. me" has dominated much of human history. It meant choosing to serve yourself or others, but not both at the same time. In reality, we are one with each other at our deepest level! We *can* act in our self-interest and the interest of others at the same time. That is made clear in the upcoming section on "selfness".

It's right to empower people and prevent exploitation of them by those with political power. Some influential advocates of the philosophy of self-interest have put their reputations and lives at risk, in an effort to educate people about the horrors of communism in the hands of brutal dictators. The irony is that such people were also being altruistic in that sense, when viewed in a larger context. Sure, they profited from books sales. However, that resulted from attempts to protect *others* from cruel governments.

According to California's UC Berkeley researchers, "Recent neuroscience studies have shown that when people behave in an altruistic way, their brains activate in regions that signal pleasure and reward, similar to when they eat chocolate (or have sex)." That's just one of many cases where incentives for actions have a basis in both self-interest and altruism.

Seeking Highest and Best, Not Lowest and Worst

Another irony is even far more tragic. During Stalin's 30-year rule, from 1928-1953, an estimated 20 million Soviet people died as a result of his actions. By comparison, Germany invaded the Soviet Union in 1941 and up to 27 million* Soviet people died from all causes related to WWII. (*Exact figures are disputed.)

What's the tragic irony? Many leaders of American finance and industry are strong advocates of American capitalism. However, an insidious abuse of capitalism by some of America's leading business people provided Hitler with strategic support. He used it to devastate Europe, the Soviet Union, and elsewhere. Without aviation fuel supplied to Hitler by William Stamps Farish, The CEO of Standard Oil of New Jersey, Nazi Germany would not have had the Luftwaffe military air force in full operation. The Luftwaffe was able to destroy most of the Soviet Union's 20,000 air force planes before they took flight.

Do you think today's advocates of self-interest would approve of the free market capitalism exploited by the American capitalists who gave, or sold, Hitler support? They wouldn't, at least if they abided by basic human morals which condemn initiated force.

Those business people might have said, "Hey, it's not personal, it's just business. I have people to pay and mouths to feed. I'm doing this in my own self-interest." They could have tried to justify their actions by selecting limited aspects of the philosophy of capitalism. They might have thought they were doing the right thing within the context of their small sphere of influence, sense of self, and the so-called virtues of self-interest.

Every philosophy and system of government has potential for highest and best use, or lowest and worst abuse. Each of us applies our best skills to do the same thing, with different levels of success.

I focused on free-market capitalism and its basis in self-interest philosophy because it represents a pervasive and unhealthy mindset in our world. The vital aspect of heart is a missing ingredient in this philosophy and the mindset it supports.

Heart-related behaviors such as kindness, compassion, empathy, and forgiveness are absent, or underemphasized, in narrow uses of self-interest-based philosophy. That's not a small omission. As Aristotle wrote:

"Educating the mind without educating the heart is no education at all."

A metaphor from quantum physics is useful here. Did you think we'd get out of the Attic of Outdated Ideas so soon? Well, almost, but first there's something a little weird to discuss. It used to be that classical physicists understood electrons as particles, like little balls bouncing around. Then quantum physicists discovered both light and electrons as not only particles but also like waves rolling across water.

Even stranger is that they could be both particles and waves at the same time. That's the theory of "particle-wave duality." It still mystifies physicists even though most accept it as true. Another mystery is that if you *look* at an electron, it can become either a particle or a wave. Your consciousness, or mind, can affect matter.

One function of the mind is intellect, which finds *differences* between things in an *objective* and exclusive way. That aspect of mind analyzes differences in physical properties, ideas, knowledge, behaviors, and the diversity of reality, which includes subatomic particles. The intellect is similar to the particle aspect of an electron with its distinctive features. The heart feels and finds *similarities* between things, in a *subjective*, inclusive way. It synthesizes, harmonizes and holds things together on the level of feeling. It feels LOVE. It's like the wave function of electrons described in physics.

A philosophy that emphasizes only mind, intellect, and objectivity is incomplete. It's similar to a physicist who wants to work with only particle aspects of an electron and reject wave aspects. Without knowing wave qualities of an electron, complete knowledge of its reality remains unknown. Just as an electron can be a particle and a wave at the same time, we now also know that someone can be of service to their self and others at the same time. It's not *you–me duality*, though, because it's based on *unity*!

Leaving heart out of *any* philosophy is a major flaw. It causes big and small problems. The heart unifies, harmonizes, expands, and is compassionate. It can recognize value in others, and connections with others. Later chapters of *RAISE the ZONE* will reveal an antidote to this missing part of the philosophy of free market capitalism that is based in narrow self-interest.

Transforming our problem-based society into a solution-based civilization demands work from our side. We need to develop, balance, and integrate both heart and mind. Our attic visit will end by acknowledging a particle/wave expert who also highly valued a balance of mind and heart.

Albert Einstein expressed his view that human thoughts and feelings were part of the wholeness of nature and the universe. He viewed our experience of separation from that wholeness as a kind of "optical delusion." He wisely encouraged humans to widen their "circle of compassion" to enhance wholeness of their nature.

Enlightened is the Key

Enlightened is a key word that distinguishes enlightened self-interest from greed and selfishness. People who care mostly about preventing government interference, don't seem to care as much about how businesses undermine and interfere with competitors.

Enlightened self-interest is vital for our morality, health, and honest prosperity. You could view it as a "win-win-win" situation where you, I, and everyone else always win. Without that, honest capitalism loses its fair market and level playing field that inspire innovation and productivity. Without enlightened self-interest, wealth and abundance are awarded to some, while most others are punished and oppressed.

LOVE and coherence share some major characteristics with the word *enlightened*. What makes one person more enlightened and selfless than someone else, or more than they were before? The main distinction between selfish and selfless (altruistic) depends on how you to define "self." A selfish act to one person might be altruistic to someone else. The word "self" with lower case "s" is what most people identify with as their personal life. It includes our body, appearance, feelings, thoughts, beliefs, emotions, desires, actions, lifestyle, possessions, and so on. It's all about "me," and "my" life. Whatever people do to support any of those aspects of themselves is an act of self-interest.

There are also people with more expanded awareness. They experience deeper harmony, coherence, unity, and oneness with more expanded concepts of self. To them, *self* includes friends, family, co-workers, and neighbors. They identify these as their self and their life. They see all of these people as extensions of themselves. They have a bigger sphere of awareness and sense of self.

What they do for these others, they're also doing for their own self. In that sense, they're also acting out of self-interest. To someone who doesn't have that expanded concept of self, people that behave in their family's and community's interest would seem to be more selfless and altruistic. In which category do *you* fit?

The distinction between selfish and selfless depends on how one defines "self."

Sliding Scale of Selfish vs. Selfless

What's the dividing line between selfish and selfless? It depends on who perceives the boundaries of the sphere of identity. A person who helps his or her neighbor might not have a sphere of identity and sense of "self" that extends to people in foreign countries, or other religions and races. They might mistreat or disrespect these other people and still view themselves as altruistic.

Consider someone whose sense of "self" expanded so much that they feel harmony, unity, and oneness with all of life on Earth. They would treat those with very different origins and beliefs as an extension of their "self." Their sphere of identify and greatly expanded sense of self embraces the self of everyone.

John Donne's poem, "No Man is an Island," is a perfect example of this more expanded sense of self. The parts of his poem that move me the most are, "No man is an Island," and "any man's death diminishes me, because I am involved in Mankind." Donne identified with a greatly expanded sphere of self.

If you felt such a high degree of unity with every person, you wouldn't harm other people any more than you would harm yourself. What you do for any others is done for yourself as well.

You wouldn't dump toxic waste into a foreign country and put people at risk, any more than you would dump it in your backyard. You wouldn't harm magnificent creatures of our world any more than you would shoot your pet for sport.

Whatever expands anyone's sense of self will shift them from perceived *selfishness* to more *selflessness* on the spectrum of their behavior and motivation. In each degree of expansion or contraction of the sphere of self, you could view actions of yourself or others as either selfish or selfless. Again, it depends on how expanded your sense of self is.

Mystics, saints, scientists, poets, athletes, astronauts, and people hiking in nature have shared their insights from experiencing life-changing expansion of their self. Such experiences can happen to anyone for a moment, a few minutes, or a lifetime.

Einstein, Astronaut Edgar Mitchell, Black Elk, and others have shared their view about the unity that human beings have with each other and the universe. Poetry, science, philosophy, art, and every other kind of human exploration have described such higher realities and states of consciousness.

Extreme mental clarity and elation often accompany such feelings of vast connectedness. Those experiences seem to have no limit when combined with a transcendental level of universal peace, LOVE, and harmony with one's environment. Some people are born with it, or it happens to them spontaneously one or more times for no apparent reason. However, it can also be cultivated through various means, often involving some form of regular practice of meditation or prayer. It can become more permanent too, instead of happening for only a fleeting moment.

Has this ever happened to you? Is it something you share with your friends? When it occurs, it creates an opportunity for everyone to keep expanding their sphere of identity and connectedness to an expanded environment. As that happens, the line separating selfishness from altruism continues to shift or even dissolve. How far can this experience go? Let's look as some descriptions to help us grasp this phenomenon.

Cosmic Self-Identity

John Hagelin, Ph.D., is a Harvard-trained, world-renowned quantum physicist. Dr. Hagelin has pioneered using unified field-based technologies that are proven to reduce terrorism, crime, war, and violence of all kinds. His approach to promoting peace throughout society is with methods derived from ancient Vedic sciences of consciousness.

He has often quoted Rik Veda 7.1.1, "Dure drisham grihapatim atharyum." The translation is, "At a far distance is seen the in-dweller of the house, reverberating." Dr. Hagelin has explained that experience as "reverberations" and a "cosmic blow-up of our innermost Self" that one sees at a far distance.

That means the sphere of our identity of self could be infinite and extend to the whole universe. It's not a common experience, but is still possible to achieve. That degree of expansion of our concept of "self" transcends all time, space, and other aspects of our manifest life. It's no longer limited to our personal traits, thoughts, desires, friends, family, communities, causes, or even humanity. It's a universal and cosmic sense of self.

Rising to Selfness

This cosmic experience demonstrates the highest and most expanded level of consciousness. That's why it would be "Self" with a capital "S." In that exalted state, one thinks and acts in full accord with, and support of, intelligence that creates and maintains all of creation. That's the ultimate degree of altruism. Yet, what one does for all of creation would still be an act of Self-interest. It's just that one's sphere of identity has expanded to an infinite cosmic level.

At that level, any remaining sense of duality rests on a foundation of unity and oneness. The sense of *me* no longer dominates my identity compared to anyone or anything that is not me. It's total unity with the essence of who, and what, we are. It's the same essence of who, and what, everyone or everything else is.

The definition of "Selfness" is when we experience our "Self" in its most expanded state. That's our highest level of human development. Debates or conflicts between self-interest vs. altruism could not apply to any person at that level. More importantly, along with all of that expansion of "Self" identity comes more happiness, joy, LOVE, and a feeling of infinite abundance of life.

That's what increasingly *raising the zone* means on a personal level. It's not an all or nothing experience, though. It's more often a gradual integration of higher frequencies that result from expressions of LOVE, truth, and coherent brain waves. John Donne could have also written in his poem, "Any man's joy enriches me because I am one with mankind." Everyone has this capacity. We also raise the zone for others when we do it for ourselves.

Expanding Selfness of the Masses

Is there any way to expand the spheres of self of the masses toward more enlightened self-interest? Can you accelerate the pace at which people gain a much larger self-identity? Yes, to both questions, and we'll get to that soon. Doing so would provide greater support to the emerging trends that Dr. Aburdene has identified. She also talks about raising consciousness through spiritual practices within businesses and personal life.

Raising the Zone is about pushing the right buttons to increase the enlightened part of self-interest. It's about expanding LOVE and coherence. These are the keys for us to evolve more quickly into a solution-based economy and human condition.

Extraordinary practical strategies are already laying a firm foundation for a more enlightened, solution-based civilization. Buckminster Fuller urged people to "build a new model" when they want to change something, and not try to fight the old model of reality. Enlightened individuals, companies, organizations, philanthropists, and entrepreneurs are building this new model and paradigm all over the world. Our spheres of influence overlap in all directions. An ocean of coherent spheres can lift our world!

Inspiring People to Stay Alive and Thrive

Inspirational and motivational books, articles, films and other media are helping to usher in a solution-based civilization. Those learning and applying this kind of information can spread these benefits to uplift others even more. These books have transformed, improved, and saved lives of people all over the world.

It's impressive enough that Jack Canfield and Mark Victor Hansen created extraordinary success with their series of *Chicken Soup* books. Canfield shared one statistic that I found to be the most inspiring part of their story. It wasn't about book sales, revenue, fame, or fortune. He said that they received letters from over 2,000 young people who thanked the authors for creating books that inspired young people to not commit suicide.

Many of us know someone who has taken their life. It's so sad and tragic that anyone could suffer so much to think life was not worth living. I remember the sad night when one of my childhood friends ended his life in the summer of 1969. We lived just a few houses away from each other.

What could have given them the buoyancy to rise above the depths of their anguish? What could have convinced them that their life was special and a better future was ahead? We all wish we could have done something to prevent such tragedies. We all have something inspiring to share with others. If you know of someone who is suffering and in deep despair, reach out to them, because you might prevent a tragedy. There are actions, even at a distance, to help them fill up their sphere of identity "life-jacket" and keep them afloat. It can keep them above the dark and turbulent waves of their agony until they regain their self-worth and inspiration to continuing living.

We are all on the same "spaceship Earth" as Buckminster Fuller called it. We all fill the ocean of collective consciousness with frequencies that either raise or lower the tide of life. Building a broader base of enlightened self-interest will raise all of our boats on that ocean to open greater opportunities.

Summary

The most important foundation for any person or society is LOVE and coherence and the balanced education of both mind and heart. That will expand our sphere of identity and sense of "self" to a state of more enlightened self-interest. The result will be a merging of self-interest and altruism.

Flawed notions in free market philosophy can be overcome by raising the zone of our individual and collective consciousness. That will achieve the highest goals of economic and governmental systems.

Many approaches are building momentum towards a solution-based civilization. The more we collectively expand our spheres of "self", the more doors will open to an amazing new Garden of Opportunities.

Various ways exist to expand your sphere of identity to realize your highest "Self". You can also indirectly help expand the spheres of identity and sense of self for others when they are struggling to do it themselves, or don't even know why it would help them. The following chapter provides an understanding of how truth, coherence, and LOVE are our new keys to an urgent global transformation. Now that we've found what we needed, are you ready to leave the Attic of Outdated Ideas?

10

Truth, Coherence and LOVE – Antidotes to Stress

"Enlighten the people generally, and tyranny and oppressions of body & mind will vanish like evil spirits at the dawn of day."
— Thomas Jefferson

As we continue on our tour, would you enjoy some unique "refreshments" for your mind, heart and soul? Does that sound like a clue that we're going to get more philosophical and even scientific for a few minutes? Jefferson talked about enlightening the people generally. Nikola Tesla regarded energy, frequency, and vibration, as secrets of the universe. Are you now wondering why a chapter on truth, coherence, and LOVE would include one of America's founders and legendary inventor Nikola Tesla?

Imagine we're on a mountaintop. All that we see is our natural environment. Puffy white clouds drift slowly across an azure sky. Spectacular mountains, valleys, waterfalls, lakes, and wildlife surround us. Majestic beauty uplifts us. We're humbled with a sense of peace, gratitude, and wonder.

Truth, coherence, and LOVE are closely related and share common features including harmonic frequencies, or vibrations. However, this chapter will not reduce these concepts to an emotionless laboratory experiment. It merely begins a discussion of human-based technologies that are essential for our survival. It reveals harmony as a key technology for our urgent transformation.

This discussion won't diminish ineffable human feelings of romantic love either. It won't attempt to replace that sublime and cherished experience as represented through art, literature, and music throughout much of human history.

Love and harmony combine,
And round our souls entwine
While thy branches mix with mine,
And our roots together join.
"Love and Harmony" by William Blake

William Blake's poem described personal and sweet qualities of romantic love. It could also refer to another cosmic reality. At the deepest level of infinitely branched and entwined diversity in creation, we're all one at our root. Physicists would describe how we arise from a single unified field at the source of all of force and matter fields.

Philosophers might say that's where we join with our highest truth, most expanded coherent consciousness, and most profound and fulfilling LOVE. Musicians might sing that's where we exist in a oneness of perfect synchrony and accord with all others. We recognize that sublime oneness when we connect with someone through simple acts of compassion, giving, kindness, and respect.

Harmony of Truth, Coherence, and LOVE

There's a lesser-known aspect of LOVE that needs a voice and description. This chapter highlights *harmony* as an element common to truth, coherence, and LOVE. It can be a powerful and positive technology for transforming our problem-based society into a solution-based civilization.

Truth, coherence, and LOVE are inseparable. These three qualities of ultimate reality are described in Sanskrit as, "Sat Chit Ananda". One translation is, "Truth, Consciousness, Bliss." We can obviously equate truth with truth, then coherence with consciousness, and LOVE with bliss. Think of *wedded* bliss.

Chapter Two discussed how to know truth and why that's so important. It clarified Occam's razor and introduced a scientific understanding of intuition. Truth is a common thread of LOVE and coherence. Truth means something that is in accord with, or in *harmony* with, reality. Truth can be a fact, or a discovered law of nature, as in scientific truth. Scientists are still seeking an ultimate truth in a grand unified field theory.

Truing Our Self to Align with the Universe

The word "true" is also a verb or action word. To "true" a wheel is to bring it into exact alignment for the most smooth and stable rotation. We could say that to "true" human beings in their "wheel of life" would bring them into alignment with their highest Self, purpose, and oneness with creation. Seeking highest truths is how we find our connection to everything else and ourselves.

Coherence is one feature of consciousness. Pure consciousness is perfectly coherent. The more orderly or harmonious a person's brainwaves are in waking state, the more conscious they are. This improves functions of mind, body, emotions, and behavior. All of these lead to a more fulfilling life.

Love and harmony go together, as with Blake's poem. We can feel a high degree of synchrony, alignment, or being "on the same wavelength" with a person, place, or thing we love. It's like some melodic symphony playing in our heart, compared to noisy and disharmonious feelings in a broken relationship. Harmonious relationships are more fulfilling and stable because they reinforce mutual waves of trust, respect, loyalty, and affection.

Understanding the Science of Harmony

Working together in harmony results in more success. When people, or even parts of machinery, work against each other in disharmony, it increases failure. When our mind, heart, and body are in synchrony, we run more smoothly. Every aspect of our human existence benefits from increased LOVE, harmony, and coherence.

Science describes harmony in terms of wave motion. A basic physics lesson is about interactions of waves in water. When waves are in harmony and *in-phase,* then constructive interference occurs and they build on each other. When waves aren't fully in harmony, destructive interference occurs and they quickly cancel or disrupt each other.

Waves can exist in fields as energy, frequencies, and vibration. Harmonic frequencies are present in music, electronics, and other phenomena. Tuning forks and an opera singer shattering crystal goblets are familiar examples. Tesla was masterful in applying laws of nature that govern harmonic frequencies. When we feel in tune, in-sync, aligned, or resonating with someone or something, it's the work of constructive interference. It's resonance among such harmonic frequencies or vibrations.

"The Law of Attraction" is a basic tenant in New Thought philosophy. It's means that vibrations of thoughts and emotions can attract similar vibrations of energy from others and the universe. Renowned poet, Jelaluddin Rumi, expressed that philosophy as, "What you seek is seeking you." That describes Tesla's secrets of energy, frequency, and vibration.

Using Harmony to Get Unstuck

The "car-stuck-in-mud" analogy is applicable here. Have you ever helped push someone's car out of mud or snow? If so, you probably used constructive interference. You pushed forward as the other person inside stepped lightly on their accelerator. The car rolled forward a bit and then backward. As it reaches its backward limit, it started rolling forward again with some momentum.

At just the right moment, when their car bounced forward, you gave another push as they stepped on their accelerator again. That usually goes on a few times. Each time, a combination of pressing the accelerator, and perfectly timed pushing, built up even more momentum. Finally, their car rolled forward enough to be free from the mud or snow.

This story demonstrates how coordinated and harmonious waves of effort reinforced each other in a constructive way. Working in-sync, in-phase, and harmoniously was your secret. Had you pushed on their car just as it was rolling back toward you, it would have demonstrated destructive interference and a wasted effort.

Between being stuck and not stuck is the difference between having harmony or not. If mind, heart, body, and spirit work together, all aspects of health and living are better. Working with others to find common ground, and finding harmony with our environment and humanity, expands our sense of self and happiness. Even better, it guarantees our survival as a species. Our planet is stuck in a muddy, obsolete, and unhealthy paradigm. By applying a technology of harmony through truth, coherence, and LOVE, we can get ourselves and civilization *unstuck*.

Three Harmonies for Better Life

The more *truth* we know, the more harmonious we become with bigger realities of life, nature, and the universe. The greater *coherence* we create and apply, the more productive we become and the more harmoniously we progress on an individual and societal level. The more *LOVE* we experience and express, the more harmony we feel with others and ourselves.

Modern medicine has specific clinical words and criteria for defining fear, stress, anxiety, depression, and related disorders. I will use simpler terms. All of these disorders make us feel bad in some way. Acute and chronic stress disrupts our inner harmony and contributes to each disorder, which causes more stress. It's a vicious cycle and a downward spiral.

Fear of the unknown is stressful. Basing choices on lies is also stressful. Individuals, organizations, and corporations spread misinformation to profit from our ignorance of truth. They say that horrible diseases are incurable while suppressing cures for such diseases, or making cures illegal. They say that fossil fuel-based energy is necessary and the most affordable, even though cleaner and more affordable energy alternatives remain hidden, destroyed, or ignored. They alarm us with non-existent threats to keep us in a state of fear so that we give up rights and freedoms in order to be *protected*.

Truth of what was unknown sets you free. It brings knowledge and a light that dispels the darkness of ignorance. Therefore, *truth* is one key antidote to stress. Research has found that expressions of *LOVE* reduce stress. *Coherence* in brain waves is associated with less stress as well. Technologies and programs that reduce stress also increase coherence in your brain, and between your brain and heart. Increasing coherence in your mind, heart, and body, also increases resilience to excessive stress.

Harmony as a Solution Technology

The element of harmony in truth, coherence, and LOVE can now become a powerful technology. It can raise individual life and humanity above problems that plague us. Harmony is a means to get us unstuck from the mud we have created and into which we keep sinking deeper.

Greater truth, coherence, and LOVE create more happiness, health, better relationships, and success in life. Together, all three create an upward spiral of solutions for virtually all problems.

Following is a summary of how these three elements reinforce each other. They are almost interchangeable. Pause and ponder their significance in your life. Connect the dots!

- **Increased LOVE reduces stress.**
- **Reduced stress increases LOVE.**
- **Increased truth reduces fear and stress.**
- **Reduced fear and stress open us to greater truth.**
- **Increased coherence reduces stress and disharmony with others.**
- **Reduced disharmony with others increases LOVE, truth, and coherence in relationships.**

Intelligence of the Heart

HeartMath Institute Researchers share science-based beliefs that human intelligence is mostly "heart intelligence." Their book, *Heart Intelligence,* introduces the idea that focused LOVE is a powerful means to help people manifest their "undiscovered gifts and talents" and get beyond fear and its effects.

Nikola Tesla talked about how *instinct* can transcend knowledge. He mentioned, "certain finer fibers" that give us an ability to "perceive truths when logical deduction...is futile." Could the "fibers" mentioned by Tesla equate with finer feelings related to intuition and "heart intelligence"? Intuition is our ability to know something instantly, from our "gut feeling," without going through the process of logic and reasoning. Heart intelligence, intuition, and instinct, are closely related.

Heart intelligence has more functions than mere intuition. HeartMath researchers have measured physiological coherence derived from patterns within heart rhythm variability (HRV). They've seen orderly and harmonious synchronization of the heart, respiratory system, blood-pressure rhythms, and other bodily systems. Through one HeartMath technique, I can reduce my systolic blood pressure 15-20 points in back-to-back measurements. This kind of coherence results especially from positive emotions. The next chapter shares more details on that.

An intellectual approach of the mind, and an emotional approach of the gut, or heart, are two parallel paths to finding truth. Heart-based approaches to finding truth seem deeper and more profound among these two contrasting approaches.

"Harmony" and "transcendence" are common elements too. HeartMath's "undiscovered gifts" line up with Tesla's "secrets of the universe." Transcendental abilities of the heart reveal gifts and secrets. We find ultimate truth, coherence, and LOVE at that transcendent level of our roots, described by William Blake. Let's now look more at how LOVE and coherence reduce stress—one of the biggest obstacles to becoming a solution-based civilization.

Jefferson's Enlightened Prophecy

Do Jefferson's words, quoted at the beginning of this chapter, still matter in our modern age? Yes, and it might be one of the most insightful and prophetic statements in history. Modern science validates the benefits he described from enlightening the people. His words could also be: "Create LOVE and coherence among the people generally, and tyranny and oppressions of mind, body, behavior and society will vanish like evil spirits at the dawn of day."

Why are "the People" not enlightened anyway? Why aren't we free from oppressions of body and mind in the first place? Stress is the main cause. It creates chaos instead of coherence. Let's review the effects of stress on our mind, emotions, body, and society.

How Stress Amplifies Weaknesses

A lesson from basic physics illustrates how stress amplifies weaknesses. But first, here's two descriptions of stress:

1. One kind of stress is the exertion of pressure on any kind of material, whether on a bone, window, metal, etc.
2. Another type of stress is what you feel when under too much pressure of time, workload, expectations, pollution of various kinds, health problems, political chaos, etc.

During my one year at Maharishi International University, an example from physics stuck with me to this day. It was a lesson on how stress amplified defects in a glass object. First, a glass rod was supported at both ends. When the glass was pure and free from defects, it held a lot of weight until the glass finally broke. The photographed stress lines were mostly uniform, parallel lines running through the glass from end to end with no disruption.

If the bottom side of the glass rod received even a slight scratch, something different happened. The stress line closest to the scratch made a tiny blip as it formed around the scratch. The next line above went around that blip with an even bigger blip.

Each line above also went around each line below it, increasing the blip's size. The lines looked, and acted, like a wedge driven into the glass. Stress lines gathered around the scratch, amplified its weakness, and caused it to break under far less weight.

Abundant research on stress reveals its effects on human life. Excess stress causes or amplifies a majority of mental, physical, behavioral, and societal problems. Most people are more vulnerable to such problems when they're under too much stress.

Tour Exercise - Take a moment to ponder these questions:

- Has stress amplified weak points in your life?
- Did you become angrier and more irritable?
- Did your memory or focus decline?
- Did you develop heart disease, worsening asthma, indigestion, or a weaker immune system?
- Did you become more aggressive or enjoy life less?
- What did you do to overcome the excess stress?

Stress Triggers Unnecessary Genetic Expression

Stress amplifies genetic predispositions to disorders in health and behavior. "How Stress Changes Your Genes," an article in *Psychology Today* on January 6, 2015, addressed this. Mitchell L. Gaynor, M.D., author of *Your Genetic Destiny,* wrote the article. Dr. Gaynor describes how anxiety negatively affects your genes. He offers advice to keep it under control.

Gaynor claimed that different sources of stress, including unhealthy diet, and personal habits such as smoking, affect genetic expressions. He describes a field of study called *epigenetics* that explores causes of unnecessary expression of genetic tendencies.

Dr. Gaynor explains that specific gene mutations, such as those that predispose women to breast or ovarian cancer, can be "turned on" by excessive stress when they would not turn on otherwise. He says their destiny is not fixed. Dr. Gaynor cites research published in *Proceedings of the National Academy of Sciences* in 2013.

Apparently, chronic stress can result in unnecessary genetic expression of a range of health problems. Among well-known therapies that Dr. Gaynor suggests to reduce stress are meditation, exercise, and relaxing music. All of these have extensive research showing specific mental and physical benefits that are opposite of the effects of too much stress. Sex can be an effective stress-reliever as well. Moreover, reducing stress can improve performance and increase sexual enjoyment.

Environmental Stress

Air pollution is one major source of stress. Electro-pollution from cell phone radiation, "smart meters", building wiring, and computers are also sources of "electro-stress" for the masses. It creates a M.E.S.S. (Mass Electromagnetic Stress Syndrome). Some activists have declared that 5G technologies will create even greater increases in various mental, physical, and behavioral disorders such as autism, Alzheimer's, ADHD, asthma, autoimmune disease, brain tumors, and diabetes.

When 5G cell towers invade communities, schools, and government buildings, more health problems will emerge. It's another health hazard that politicians, telecommunications companies, and bureaucrats will claim is safe, despite many personal experiences and scientific research to the contrary.

Scientific American's article on May 27, 2016, announced findings by government scientists that presented strong evidence of a connection between exposure to wireless devices and cancer. The article stated, "The International Agency for Research on Cancer in 2011 classified RF radiation as a possible human carcinogen."

Studies on stress from wireless technologies would be more useful if they compared typical *chaotic* radiation of cell phones, with *coherent* radiation from technology that reduces signal "noise". Chapter 12 explains this more. Preventing harm from electromagnetic pollution could be extremely practical and affordable, even for entire metropolitan areas to clean up their M.E.S.S!

Stress Causes "Swiss Cheese" Brains

Dr. Daniel Gregory Amen has shown that severe brain damage results from acute and chronic stress. Images of living brains in victims of extreme trauma from physical, mental, and emotional abuse or substance overuse, look much like Swiss cheese. "Holes" appear as areas of weaker metabolic brain activity. Dr. Amen is a brain disorder specialist and American psychiatrist. As the director of Amen Clinics, he offers ways to reverse such brain damage.

The prefrontal cortex is a critical brain section. It's considered the brain's CEO. Damage to that area reduces higher cognitive functions such as ethics, decision-making, and conscience. Such damage reduces a person's ability to filter harmful and impulsive behavior, and to comprehend results of ones' actions.

Collective Stress Affects Communities

Using the glass rod analogy, imagine stress lines gathering around weak or vulnerable members of a community. Their stress magnifies their weakness and causes them to "snap". That could cause a single event such as a mass killing, or a series of snaps as with repeat violent offenders, drug users, and child abusers. It causes snaps in ethics for white-collar criminals. It could increase racism, bullying, cruelty, and general disrespect for others.

That was mostly my reason for creating the *Safety through Peace* ballot initiative. Reducing societal stress can reduce stress-related problems across a full range of social indicators. It doesn't affect everyone equally. Higher stress levels in cities impacts these weaker links and results in higher levels of crime, substance abuse, violence, suicides, homelessness, and other societal problems.

Collective societal stress puts "holes" in collective ethics, respect, harmony, compassion, and happiness of a community. This chaos of societal stress is a "virus of incoherence". It harms everyone to some degree. Chronic societal stress causes a rapid degradation of society, which in turn causes mass degeneration of government. However, societal stress can be reduced.

Swimming in Our Ocean of Consciousness

Take a moment to view collective consciousness as an ocean in which we all live. In a calm ocean, where waves roll along in a more coherent flow, you can see far. You can feel a more peaceful connection to expanded horizons of water, even as waves gently rise and fall. It's more peaceful. At night, you can feel a much closer connection to the universe under a sky full of stars.

In a turbulent and stormy ocean, you would likely feel more anxiety and fear. Your vision goes only as far as the next random crashing wave. You feel far less connected to, and in harmony with, the vast ocean. You would barely notice stars above you, even if you could see them. Instead, you feel threatened by each wave that might bring your life to an abrupt end.

That's the difference between coherence and chaos inside each person and in collective consciousness. In such turbulent waves of our incoherent consciousness, each crashing wave threatens our existence and ability to share and care. That urgency of survival clouds our perception. Our desires meet resistance, and conflict more with the desires of other people who are also clouded by stress and chaos.

Jefferson's Confidence in Enlightened People

Jefferson's words above were prophetic, but not because of his intent when writing those words in a letter to Pierre Samuel DuPont de Nemours, dated April 24, 1816. His intent then was to promote general education to the masses. Jefferson urged that "the People", from whom our government's power is derived, should be informed. However, there's a deeper meaning too.

"Wherever the people are well-informed, they can be trusted with their own government."

—Thomas Jefferson

Jefferson proposed studying subjects including: anatomy, astronomy, botany, chemistry, languages, mathematics, philosophy, natural history, simple diet, and rigorous exercise. The courses he suggested have been standard studies in general education for decades. However, it's a stretch to claim that they have caused "tyranny and oppressions of body & mind" to "vanish like evil spirits at the dawn of day." Violence, bullying, racism, drugs, and drinking to "the dawn of day" plague schools and colleges.

Stress is the leading cause of these problems for students. The misuse of knowledge proposed by Jefferson has caused wars, disease, and problems for the masses to benefit the few. Studying such common academics is not enough to achieve the aims of his insightful declaration.

Jefferson's use of "enlighten" is intriguing. He could have used "educate." Definitions of "enlighten" include imparting both intellectual and spiritual knowledge, and insight. They also include informing, briefing, or making someone aware of something. The word "enlighten" also has higher meanings that originated long before Jefferson's time.

"Know thyself" is an ancient Greek aphorism. Its meaning and source are still debated. Marianne Williamson, a spiritual teacher, author, and lecturer, said that, "Self-awareness is not self-centeredness, and spirituality is not narcissism – 'Know thyself' is not a narcissistic pursuit." Many spiritual traditions view a higher transcendental "Self" as above and different from, yet inclusive of, the "self," which equates more with individual ego. "Self" is known as pure coherent consciousness, infinite creative intelligence, bliss, unbounded pure awareness, and LOVE.

These attributes describe the ultimate source of every sentient being in our universe. Knowing the Self is an experience, and not merely an intellectual exercise. However, various literary works attempt to describe it in all of its unfathomable expansiveness and richness. Physicists such as Dr. John Hagelin would explain "Self" as the unified field of all the laws of nature. He also would describe the unified field through mathematical equations.

The "Age of Enlightenment" was an intellectual movement in Europe during the seventeenth and eighteenth centuries. It emphasized human reason, science, toleration, and the possibility of creating less tyrannical governments. Jefferson's quotation relates to these areas. However, it applies even more to spiritual enlightenment through higher states of consciousness.

Renowned saints, yogis, mystics, and ordinary human beings, who engage in powerful spiritual techniques, have attained these higher states. They also knew the power of LOVE. Those ultimate states of enlightenment are characterized by highly coherent brain function and mind-body connection.

This experience can cause "tyranny and oppressions of body & mind to vanish..." even for short periods. Experiencing more coherent brain waves can reset, or rewire, your brain to create higher consciousness. Truth about enlightenment and freedom from stress has existed for a very long time. However, merely accepting that such truth exists isn't enough. You have to want truth and LOVE to recognize and be receptive to them.

LOVE and coherence enable us to be more sensitive to what is true or false. Less brain damage from stress, and increased brainwave coherence, would reduce lying. More coherent wakefulness and clarity would reduce the tendency of people to accept lies without question. Scientific research has shown a broad spectrum of benefits from attaining the highly coherent state of "Knowing thy Self." For example, hundreds of studies on the daily practice of Transcendental Meditation have shown:

Body – Reductions in: stress, substance abuse, heart disease, bronchial asthma, physiological aging, and physical fatigue; normalization of blood pressure and cholesterol

Mind – Reductions in: insomnia, anxiety, mental illness, and mental fatigue; improved mental clarity, memory and general intelligence; improved academic performance, cognitive development and learning ability; greater creativity, and increased blood flow to the brain

Research on Transcendental Meditation also shows that brain coherence increases between all different brainwave frequencies, and between both brain hemispheres. Being enlightened, by "Knowing thy Self," has shown benefits to behavior as well. That would be a natural extension to vanishing "tyranny and oppressions of body & mind." Other research on Transcendental Meditation shows:

> **Behavior** - Increased friendliness; better relations with co-workers; increased job satisfaction and performance; and decreased violence and hostility.

Another aphorism reinforces this theme. In Patanjali's Yoga Sutras, 2.35 it states: Ahimsa-pratisthayam tat-sanidhau vaira-tyagah. One translation is: "Where non-injury is established, in the vicinity of that, hostile tendencies are eliminated." In other words, a person in a state of coherent consciousness eliminates, or at least reduces, violence and hostility by others in their environment.

A mind in a state of non-injury or non-violence, as in a state of yoga, is in a state of coherent consciousness. People who come near such a person spontaneously and naturally do not have feelings of hostility, or at least have less. St. Francis of Assisi was apparently in a state of "non-injury." Others throughout history achieved this state as well. It's within everyone's reach, as research on Transcendental Meditation and other means of creating brain and heart coherence have demonstrated.

Rishis, Saints, and Scientists

Quantum physics offers a parallel to the idea of protecting one's self from hostility and other outside disturbances. The Meissner Effect is the ability of a superconductor to repel disruptive magnetic fields. A superconducting material contains a super coherent state of electrons. The Meissner Effect enables such an internally coherent field to repel any outside or hostile influence coming from incoherent fields. It is a protective barrier.

Scientific explanations, even when clear and concise, can be hard to grasp if you're not familiar with such language. How about another way to understand the Meissner Effect? Following is a phrase already known by many people, especially children.

**"I'm rubber. You're glue.
Anything you say bounces off of me
and sticks to you!"**

The rubber is like a person when they're calm, coherent, stable, and peaceful. In that state, they're more resilient to cruel, hostile, disruptive, or incoherent words or negativity from others. In that more coherent state, their natural systems of detoxification and immunity are also stronger. That increased strength helps protect them better against pathogens and toxins that are impossible to avoid. So, think of the "rubber" as a coherent system described by the Meissner Effect, which repels incoherent and unhealthy influences threatening it from the outside.

Jefferson did say, "Enlighten the people generally." He was thinking on a scale broad enough to affect the responsiveness and integrity of government. He intended to upgrade the public's quality of life. Anyone can create coherence and LOVE at any level, even with the push of a button. **That can also be as easy as sincerely speaking a kind word.**

We all have buttons that turn us on to LOVE, compassion, hope, harmony, and creativity. We also have buttons that turn us off, make us angry, hurt, unhappy, or uninspired. You just need to know how to push the right buttons. Any buttons that raise the zone of LOVE and coherence will benefit you the most.

Antidotes for Global Stress

Is there any evidence that a collection of individuals knowing "thy Self" or in a state of yoga and coherent consciousness could impact society? The best example of people creating societal coherence was in Washington, D.C. during the summer of 1993.

By July, 4,000 people had gathered to prove once again that a large group practicing techniques of Transcendental Meditation could reduce violent crime rates. They used advanced meditation techniques from Patanjali's Yoga Sutras to create a field of societal coherence.

The Maharishi Effect

Results of previous gatherings and published research made the outcome of D.C.'s experiment predictable. Dr. John Hagelin, a quantum physicist, was its project leader. He asked D.C. police what level of reduction in violent crime would be impressive. They responded with "20%." Consequently, Dr. Hagelin predicted a reduction of at least 20% in violent crimes during the D.C. project. The chief of police added that it would take a snowstorm in July to achieve that, as he knew the direct relationship between high temperatures and high crime rates.

Researchers collected final crime statistics after the demonstration period. The research appeared in *Social Indicators Research*, 47, 153-201. The rate of crimes tracked in the study decreased by 23%, including homicides, rapes, and assaults.

Both ancient Vedic science, and modern quantum physics, helped to explain the D.C. peace project success in reducing crime and violence once again. Tyranny and oppression of mind, body, behavior, and even the environment did not vanish entirely, but was significantly reduced. Researchers labeled this phenomenon, the "Maharishi Effect." It's a profound principle that shows how increased societal coherence results in less violence and crime.

How many other ways are there to create societal coherence and achieve similar results? As far as I know, there's no significant body of research on alternative strategies for creating a citywide effect of coherence. However, other approaches show potential and warrant serious research. Collectively, these efforts can raise humanity's quality of life through more coherent brains, healthier bodies, and more loving behavior.

Summary

Truth, coherence, and LOVE are essential keys to open the treasure of a solution-based world. They are antidotes to stress and chaos for both individual and societal levels of human life. They help inoculate us from the "virus of incoherence" that plagues humanity. They can be experienced and expressed by any person, and can be increased in the entire population of a community, city, state, nation, and our world. Their beneficial effects have been scientifically validated at all these levels.

Truth, coherence, and LOVE can be increased even by one person or a relatively small group to create positive benefits for a much larger group. These are human-based technologies. A growing number of electronic and mechanical technologies are also antidotes to stress, and can inoculate us against the virus of incoherence as well.

The following two chapters feature some approaches from both the microscopic and macroscopic levels. You'll be surprised at how easy it can be to "enlighten the people generally," as Thomas Jefferson advocated. By the way, if you see an adorable little mouse running around, don't worry, it's the extremely kind, honored guest that will be introduced in the next stage of our tour.

11
The Giving Xperiment

"A world filled with kindness is better than cheese."
—Amazing Mouse

Better Health in 15 Seconds

Now we'll explore how your health can improve in 15 seconds, simply by expressing appreciation, compassion, gratitude, and LOVE. You'll learn how these expressions can make your heart coherent, and calm your neurological and endocrine systems.

The last chapter identified the common threads of harmony between truth, coherence, and LOVE, and their roles in reducing stress. This chapter focuses even more on LOVE's scientifically measured effects. Do you recall our "minds on" expanded thinking experiments in Chapter 4? Now we'll do some "hearts on" experiments that can benefit our daily lives.

RAISE the ZONE gives you simple keys to create a new life with power that you already have. Your nature is infinite LOVE. The ultimate technology for transformation is LOVE. It's all about connecting to others and ourselves. **The challenges of life can cause us to lose sight of our simple ability to LOVE.** Science reveals why we feel better from expressions of LOVE, and the extent to which LOVE can spread to other people.

Moving Into the Light of LOVE

Our tour will now go into the sun-filled atrium. Do you feel the sun's warmth and brightness? LOVE does that too. Does it feel healing and nourishing? Being with another person, a pet, or in the presence of nature, can create that good feeling as well.

Researchers have measured health benefits of expressing LOVE. Our "hearts on" experiments are courtesy of the National Giving Xperiment (Gx) initiated by the One World Heart Project (OWHP), and founded by Cam Vuksinich. Cam is as insightful, giving, and compassionate as anyone you could ever meet. A core part of Gx is Cam's book, *Better Than Cheese.*

> "*Better Than Cheese* is a little book about a BIG idea. At first glance, it looks like a children's book but guess what, it is not just for kids. This simple book unlocks the secret hidden in each heart. We were born to love, and giving is love in action. Giving has the power to change lives forever. Decades of scientific research support the timeless, universal message in this simple book that giving is good for you.
>
> Toddlers, under the age of two, exhibit greater happiness when giving treats to others. Giving in high school predicts good physical and mental health into adulthood, a span of over 50 years. Giving reduces stress, increases happiness and longevity, and protects the giver in all stages of life."

Unlimited LOVE and Instant Health

Cam's inspiration for the Giving Xperiment came partly from the book, *Why Good Things Happen to Good People – The Exciting New Research That Proves the Link Between Doing Good and Living a Longer, Healthier, Happier Life,* by Stephen Post, Ph.D., and Jill Neimark. Gx collaborates with Dr. Post, Executive Director of the

Institute for Research on Unlimited LOVE (IRUL). Philanthropist, Sir John Templeton, selected Dr. Post to lead IRUL, which is funded generously by the Templeton Foundation.

Dr. Post's book describes ten expressions of LOVE, and highlights various benefits of practicing them. Researchers have documented many of these benefits. Expressions of "LOVE in action" in the Giving Xperiment are simple acts such as Gratitude, Compassion, Loyalty, Respect, Humor, Courage, Forgiveness, Listening, Helpfulness, and Creativity.

Dr. Post shares astonishing research about gratitude. Even the slightest intention of gratitude is potentially life changing. He says that we can shift our nervous system into a calm state just by having gratitude for five minutes. He cites a study in 1995 by Dr. Rollin McCraty, director of research at the Institute of HeartMath in Boulder Creek, California.

The study describes a correlation between states of appreciation and synchrony, or resonance, among various rhythms of the heart, breathing, brain, and other physiological functions. These gain a state of coherence in response to positive emotions.

That's opposite of the erratic effects of having stressful and negative emotions such as anger. Additionally, positive emotions reduced the stress hormone cortisol, and increased DHEA, the healthy hormone. Please remember the words "synchronized," "coherent," "erratic (chaotic)," and "stress."

In an interview with Cam in May 2016, she shared the scientific basis, purpose, and early achievements of her innovative project to spread the practice of loving behavior through various ways of giving. Similar to Dr. Post's book, she created activities involving attributes of LOVE. Gx gives special importance to nurturing these attributes in young children.

Let's learn directly from Cam about her project. True wisdom of the heart, and the magic inside everyone, will be unveiled right before your eyes. [Note: Tour Guide Suggestion – Make a list of the keys to giving that Cam shares. They are the simplest, easiest, and most affordable ways to raise the zone of your life and others.]

Giving Xperiment Discovers "Amazing Mouse"

Jeff: What's the Giving Xperiment? What would you most like world leaders and the public to know about its overarching theme?

Cam: The Giving Xperiment is a simple answer to a complex problem. Fear is a preexisting condition, and LOVE is a universal cure. We all want to love and *be* loved. In every moment, people have a choice to include LOVE in their daily diet. The Giving Xperiment is a LOVE vitamin.

LOVE is truly the greatest power. Nothing has more value, purpose and sanctity than LOVE. The only way to bring back a sense of balance is to return to the wisdom of the heart and our natural ability to connect.

When I first started this project, I wanted people to know that LOVE needs to be re-defined and re-understood. It has been misunderstood and used as a tool of manipulation and suppression. We want to elevate the understanding of LOVE as a resource that is renewable and expansive. It is vital for people to understand their active role in expanding consciousness and LOVE. It is as simple as a smile. That active part of our spirit, of consciousness, is real. It is not theoretical.

Jeff: How does the Gx play into that?

Cam: Gx returns people to their source. They do not have to wait for someone to give them a job, money, love, or respect. It returns them to the core understanding that the source is within.

Jesus said the kingdom of heaven is within. Every human being is the source of all that is infinite, all that is renewable.

It is through that flow of LOVE that abundance grows. LOVE becomes stagnate in a closed system. Once you introduce a drop of LOVE, it becomes inclusive and grows. Open systems do that. LOVE is not afraid to include new ideas. Each one of these qualities in the Gx values and virtues goes all the way back to the Greek philosophers. They talked about the ideals of human behavior.

We are craving the love connection. Without that experience, we seek to satisfy our need through alcohol, drugs, hoarding money, and unhealthy sexual behaviors that people think will fulfill them. They believe these things will allow them, for a moment, to feel greater than the emptiness they feel. However, they merely connect to even more emptiness, which is harmful.

That's why it is crucial for us to start connecting. We disempower ourselves when we fail to connect with another human being. When we connect to others, our universal power of LOVE amplifies in magnitude simply by that single connection. It is not just the power of two connected; it amplifies and multiplies. Then other people within that more coherent field of LOVE join that stream of consciousness and expanded field of awareness. Dr. Dean Radin measured this phenomenon of "entrainment," of affecting other people by your energy field.

Over the last ten years, I've realized the value of these virtues listed by Dr. Post, whether it is gratitude, compassion, respect, etc. We need just a drop of these loving actions. We train ourselves to want these but deny them to ourselves. At a deep level, we understand that LOVE is the root of these actions. Now I hope to help people remember that LOVE is actually who and what they are.

Jeff: Gratitude, acts of kindness, and creativity seem easier to comprehend as daily things you might do or feel. LOVE is more nebulous than these other acts. How would you define LOVE?

Cam: If I try to describe a meal you have *never* tasted, I could describe it as something similar to what you *have* tasted. Everyone has experienced respect, wanting to be respected, or watching someone else get respect. Each one of these is a tangible experience that human beings feel at some level. LOVE is so elusive, and especially at this level, that no one has fully experienced it. This is a taste of the feast.

Jeff: Are there different characteristics or qualities of LOVE?

Cam: LOVE is all of these things and therefore not separate from itself. Respect is LOVE in action. Loyalty is LOVE in action.

Any relationship initiated by any of these acts is the reunion of LOVE. However, it's a way that every human being on Earth now, could have a taste of what LOVE is. Imagine one ingredient that is common to all foods, i.e. flour, but eaten in different forms. That one ingredient in all of us is LOVE.

When anyone engages in these acts of LOVE, internally, it awakens the LOVE that is already inside them and the other person. It is like a tuning fork that has been dormant. Then, it "pings" and vibrates at that same level. It creates coherence through the remembered experience. The experience needs awakening for each of us to have these felt memories of LOVE in all its flavors.

Jeff: Is there a scientific basis for Gx regarding specific mental, physical, or behavioral benefits from acts of giving?

Cam: Yes. That is in Dr. Post's work and all the work around altruism and altruistic behavior. There are decades of qualitative and quantitative research on the physiology and psychology of altruism. Loving acts change people's brain chemistry and physiology. It affects the way their body interacts with itself. It brings people back into a remembered state of wellness in a very short time. If people do these small things over time, their body resets itself to be more resistant to disease with a healthier immune system. These acts of LOVE and giving release oxytocin and dopamine – our "feel good" chemicals.

The Science of Generosity Initiative confirmed a range of benefits from altruism and giving. Many of these are in the book *The Paradox of Generosity*, by Hilary Davidson, Ph.D., and Christian Smith, Ph.D. Davidson explained that being generous increases the pleasure response in the brain.

Dacher Keltner, Ph.D., co-director of the Greater Good Science Center at UC Berkeley, has conducted other research and authored *Born to Be Good* and *The Power Paradox: How We Gain and Lose Influence*. He wrote that people live longer, healthier, and happier lives when they compassionately care for other people. According to Melanie Haiken, on the website *Experience Life*, Keltner cites studies showing an activation of the vagus nerve that occurs from of caring for others. This nerve is responsible for numerous healthy functions in the body. These are just samples of the extensive research on the benefits of giving.

Jeff: Does the harmony you feel with others from giving, create coherent brainwaves? How does *giving* create coherence on an individual level?

Cam: It creates intracellular harmony. It creates a brain wave state that is like meditation. It is conducive to more harmonious relationships in families, workplaces, and schools. Harmony is a good word for this. Researchers at the HeartMath Institute have studied brain waves and heart waves.

The Institute has found that EMF from the heart creates a field many times greater than the field produced by the brain. Simply stated, our hearts may have more influence in our environment than our brains do. Our physiology also responds to shifts in our emotions and our thoughts.

Jeff: How would giving and other acts of LOVE create such coherence for a whole city or our entire world?

Cam: HeartMath research has shown that remembering LOVE, or some peaceful experience, generates that coherent field. Imagine walking down a street and positively affecting everyone you pass by sharing your uplifting field of coherence. It's a simple spiritual practice. Everyone can be a human generator of harmony!

Jeff: There is a Sanskrit phrase. "Tat sannidhau vaira tyagah". One translation is "in the vicinity of coherence, conflicting or hostile tendencies are eliminated." Is that what you mean?

Cam: Right. Hostility cannot coexist in a coherent environment because it is a lower level of consciousness. Either it separates itself and tries to maintain its incoherence and instability, or it rises to a higher level of coherence. If it is close enough, it will naturally rise.

Jeff: How would you measure the effect? What kind of device or scientific equipment could do that? What kind of scientist?

Cam: The Institute of Noetic Sciences' (IONS) Global Consciousness Project, and HeartMath Institute's Global Coherence Initiative, both conduct research in this area. Dean Radin is a key researcher. They use random number generators (RNG) and other means to show spikes in coherence among large populations during specific events.

Jeff: Then what you are describing in each person is an upward spiral of building each other up and growing in coherence, consciousness, and LOVE.

Cam: Yes. Dr. Post introduced creativity as the last value of LOVE. Once we move through all of these expressions of LOVE, we can create a better world in which we want to live. We do this by building each other up. The reality we create is a result of the choices we make.

I want people to feel empowered and take positive action. Disempowered people are more likely to wait for others to do something, and accept the unacceptable while they wait for others to create a better life for them. In their heart, every person knows what they want and hope for. The only thing that prevents their dream from becoming a reality is their belief that they do not have the power to make it happen. Where is their power? They have given their power away to someone else, sometimes without even realizing it. Now is the time for everyone to take back their power and use it wisely in the service of LOVE. They do not need to wait for anyone's permission.

Jeff: Especially not the government. So, how much power do people actually have?

Cam: Not the government, not corporations, and not economic systems. There is a sense of equanimity when everyone realizes "Oh my gosh, I have the power." Collectively, it is beyond what we can comprehend. Control exists only when people are disempowered. People need to feel empowered through respect, gratitude, and other expressions of LOVE.

Global coherence researchers have shown connections between consciousness and matter. If we could have a national compassion day where all people imagine the word *compassion*, then random number generators globally could record what happens as a result. Spikes in coherence start even before an event because consciousness *wakes up*. A spike in the graphs means consciousness has cohered. People connect before the event through precognitive awareness, and during the event.

Jeff: Can you say more about enlightened gaming? What is a good example?

Cam: Yes. The Journey to the Wild Divine is good. The company offers a series of virtual games that connect the participant through a biofeedback device. The games teach relaxation and meditation techniques that are necessary for the gamer to navigate the virtual worlds they experience.

The games respond to the physiology of each participant. Progress is dependent on managing one's own state of consciousness. It's mesmerizing, and the training transfers to real life situations. In stressful times, traffic, deadlines, etc., I think of the techniques and automatically start relaxing and smiling. HeartMath offers emWave, a small portable biofeedback device for relaxation training. I have rarely seen someone calm down when they're told to, "calm down." This technology gives people the power to successfully manage their reactions and live a better life. It is meditation instead of medication.

Jeff: What have been the most impressive results so far? Has Gx been used in businesses, schools, hospitals, military, etc.?

Cam: A kindergarten class in Douglas County, Colorado, implemented the first Amazing Mouse/Giving Xperiment as a three-month pilot program. Amazing Mouse played a key role as the lead character in the book *Better Than Cheese*. Yes, kids learned the values from a charming little mouse and his assistant, Dana, one of the most caring and experienced teachers I've had the pleasure of meeting.

This process of reinforcing the Gx values in the children's daily lives supports not only their willingness to practice the positive behaviors in the moment, but also the desire to encourage others to try it as well. Each child became a little "Johnny Appleseed of Goodness."

The program was so successful that it expanded to all sixty-five pre-school and kindergarten classrooms in Douglas County. The children can relate to little Amazing Mouse because he is small, and so are they. He transforms the whole school through his loving heart. He says, "No matter how anyone treats me I'm not going to change the way I give."

He displays a committed attitude to this idea of LOVE. Kids take these Gx values and acts of giving home to parents. They all rise to higher actions of LOVE and coherence. For example, they learn different ways of interacting with the world using courage and creativity.

Gx plants each of its seed values in the school. Daily action and practice are encouraged. The teacher also engages parents who then discuss the Gx values with the kids. Home life begins to include choices that focus on more loving behaviors naturally. Through the enthusiasm of the parents, it is now moving out to churches and other community groups.

Jeff: So, Cam, what is your moonshot project? How much can GX realistically scale up?

Cam: Well, buckle up and get ready for the countdown! My moonshot is a National Giving Xperiment. I imagine cities and states "adopting" one of the core values*. The citizens could participate in an online poll to select the value they want their state to represent. For example, Colorado might choose Humor and invite people to visit Denver, the Smile High City. Kansas City recently created KC Nice. Smile High and KC Nice could engage in a friendly competition of goodness, like a LOVE Olympics. Schools, churches, businesses, and families could also choose a value and create their own Giving Xperiment – a Gx day.

I always imagined Gx as a prescription for the *Good Life* – a life inspired by LOVE. This is a moonshot that we could launch today. LOVE is the rocket fuel that our world desperately needs, and everyone is born with the natural ability to take flight. So, if you're interested in signing up for the National Giving Xperiment "moonshot," I am looking for ideas and support. Welcome aboard! [*Gratitude. Compassion. Helpfulness. Forgiveness. Courage. Respect. Humor. Listening. Loyalty. Creativity.]

Jeff: That is a grand vision of simple, small ways to make a big improvement in our world. Is there anything else you would like to share about creating coherence and the Giving Xperiment?

Cam: Yes. Small is the new big! People underestimate the power of small actions. They do not understand that one simple moment of LOVE connects us to everyone.

The Global Coherence Project has demonstrated that we are always connected. The Butterfly Effect is a beautiful example of how one action, like a smile, creates a ripple.

Between the virtual campaign, school programs, senior programs, and Gx for business, we plan to increase awareness of the social and health benefits of values-based altruism. Adam Grant's book, *Give and Take: A Revolutionary Approach to Success*, clearly demonstrates the benefits of a giving attitude. We are at a crossroads. People have a choice for either community or disunity. We come together, or we fall apart. LOVE is the answer. Even a tiny butterfly of LOVE can create a worldwide ripple.

That is values-based, instead of money-based, living. I always have the resource of LOVE to give. Anyone can do that any place and any time. Everyone always has the choice to be altruistic in every moment. If LOVE is an infinite resource, you can always be a giver and always have something to give. [End of interview]

Butterflies Surfing LOVE Waves

Cam's Giving Xperiment is one of the best examples of how humans can apply the Butterfly Effect to benefit their daily life.

The Butterfly Effect is what people refer to when they say, "A butterfly fluttering in Brazil can create a tornado in Texas." However, it's not necessarily a tornado. It could be a small act of LOVE that generates harmony and LOVE at a much larger scale. You could say it's butterflies surfing LOVE waves.

Chaos theory relates to the Butterfly Effect. The idea is that a small coherent effect in a large incoherent system can grow. It can become a much bigger influence over time or somewhere else in that system. If the system is a weather pattern, a butterfly flapping its wings in a little orderly vortex motion in Brazil could affect the early stage of a weather system that creates a tornado vortex later in another part of the world. Scientists use this principle to explain similar effects in other situations where a small action at an early stage could have larger results later.

Shortcuts to a Better Life from Giving

Cam's interview was transforming for me. The values of giving are shortcuts to better health and relationships, and a longer and happier life. They add value to life. That's why they are called "values." Have you found, though, that even these simple, small shortcuts sometimes don't happen as easily as you would like?

Do you sometimes just not feel up to expressing gratitude and compassion even while knowing their benefits? I experience that at times of extreme fatigue. Fortunately, there are ways to "prime the pump." Simple things like smiling and good posture uplift our mental outlook and emotional moods, according to research. According to other research, laughing is truly the "best medicine" for its ability to improve blood chemistry and the immune system.

Did any parent or grandparent ever tell you, "Don't slouch," or "Straighten up and fly right"? When I was mischievous or behaving badly, my mother would use the latter phrase. Sit or stand straighter. Smile. See if you feel better in any way. One phrase is "grin and bear it." How about "smile and share it"? If you do, others will too. Your light illuminates your environment.

The HeartMath Institute

Cam mentioned other pioneering organizations that research expressions of LOVE and coherence. They measure these effects on individuals, larger groups, and for the whole world. Following, is a little more about a few of them.

Doc Childre founded the HeartMath Institute education and research organization. Its explores and promotes ways to understand and use "heart intelligence." The effect of positive emotions and giving quickly brings coherent effects to mental, emotional, and physical conditions. Researchers at HeartMath also discovered that these coherent heart rhythms can resonate between two people. In other words, then get more in-sync. Cam talked about "entrainment" in regards to people becoming a generator of harmony and coherence for others.

The HeartMath Institute organizes an international effort to apply and study heart and brain coherence on a worldwide scale. According to HeartMath's project, the Global Coherence Initiative (GCI), they intend to demonstrate that intuitive solutions result from increasing heart connections to the extent of being able to shift global consciousness. HeartMath's emWave technology can create and measure coherence in psychological and physiological processes, the effects of which extend beyond the human body.

Gregg Braden, a bestselling author and visionary, has worked closely with HeartMath researchers. He promotes the concept of heart-brain coherence and has developed meditation practices to create coherence on demand.

Jon Kabat-Zinn is another pioneer bringing stress-reducing meditation practices into the mainstream. He's the founding Executive Director of the Center for Mindfulness in Medicine, Health Care, and Society at the University of Massachusetts Medical School. Kabat-Zinn has integrated elements from his practices of yoga and Buddhism with modern science. He is known mostly for teaching "mindfulness" as a means to reduce stress, anxiety, pain, and illness.

Moon-Walker Mitchell Finds Cosmic Connection

Ms. Vuksinich greatly praised the Institute of Noetic Sciences (IONS) founded by former astronaut, the late Dr. Edgar Mitchell – the sixth man to walk on the moon. His life-changing experiences in outer space inspired his research into connections between human consciousness, matter, and the cosmos.

The organization's website describes IONS as "a portal to psychic and intuitive tests and games, some of which were developed by IONS, and some by colleagues." IONS' Chief Scientist is Dean Radin. He blends scientific research with Internet technology to develop online games that test individuals' psi (psychic) abilities.

IONS also researches "collective consciousness," such as at the Burning Man Festival in 2012 and 2013. They measured if the collective attention of attendees watching the ceremonial "burning of the man (sculpture)" at the festival, would affect matter. They used random number generators (RNG) to test their hypothesis. RNGs are "electronic devices designed to generate maximum entropy or randomness."

They looked for a synchronicity, or non-chance fluctuation, in an RNG output close in time to the burning of the man. When 65,000 people watched the flaming structure, at that moment the IONS team found a "correlation with a rise in physical coherence, as measured by the outputs of truly random number generators."

IONS Global Consciousness Project, called the EGG Project, is an international, multidisciplinary collaboration of scientists, engineers, artists, and others. They collect data continuously from a global network of physical random number generators to monitor effects of synchronized consciousness from international events.

Microscopic to Macroscopic

All of the above are "microscopic" yet powerful ways to create LOVE and coherence. That means one act of LOVE or giving at a time, and one person at a time. More acts of LOVE performed by more people will increase the effect of raising the LOVE quotient and all of its expressions. This collective effort can have a macroscopic effect. Even people not directly engaged in loving acts can rise to higher states of coherence through entrainment. That is how you can radiate coherent fields and increase coherence in others nearby.

Small is the New Big

Beginning acts to transform and elevate our world can start with even *microscopic* acts of loving, respecting and forgiving ourselves first. "Small is the new big" as Cam likes to say. It's part of recognizing that we are LOVE itself. There are also *macroscopic* ways to create LOVE and coherence. Even one person can use them to achieve greater LOVE and coherence for a large number of other people. The next chapter will share examples of practical ways for doing this even on a global scale.

Amazing Mouse and his story of expressions of LOVE have transformed classrooms of pre-school and kindergarten children. Our world would quickly change for the better if schools everywhere used the Giving Xperiment, or similar programs.

Summary

Researchers have described ten expressions of unlimited LOVE, and documented many of their instant and long-lasting benefits for health, behavior, happiness, and success in life. These expressions of "LOVE in action" applied in the Giving Xperiment, are simple human acts such as Gratitude, Compassion, Loyalty, Respect, Humor, Courage, Forgiveness, Listening, Helpfulness and Creativity.

Injecting LOVE and coherence into any aspect of life can create an upward spiral. A cascade of benefits for health and quality of life will result, even if everyone doesn't knowingly and directly participate. It's like the "butterfly effect." Or maybe we should also call it "the Amazing Mouse effect!"

"A world filled with kindness is better than cheese!"

– Amazing Mouse

12
Global LOVE and Coherence

"The world is transformed by those who RAISE the ZONE!"
—The Solutionist

A Simple "Do-It-Yourself" Approach

Are you wondering *how* to create LOVE and coherence on a macroscopic and global scale? As your tour guide, I'm sure you are because it's clear you've been paying close attention. This chapter will explain extraordinary discoveries to achieve global coherence. The epic demonstration in Washington, D.C. is the most scientifically validated example of creating societal coherence. That strategy has shown effects worldwide as well. However, such projects require immense organizing to assemble and fund thousands of people needed to create the Maharishi Effect, at even the city level. Nevertheless, such vital efforts are underway.

Other technologies and programs could also reduce societal stress. The basic requirement would be to increase coherence in whatever affects society as a whole, such as air and water quality, and electromagnetic fields. When those three areas are polluted, they also cause stress and incoherence in collective consciousness. That causes problems for individuals and society. Chaos resulting from human-made electromagnetic fields is one of the most pervasive sources of this stress. So, that's where our tour goes next.

Electromagnetic Field Stress and Chaos

You've probably heard of scientific studies about hazards of electromagnetic fields (EMF). People are concerned that these fields create various health disorders and disease in living systems. They would be right in making that assumption, but not for the reason they usually assume.

All living systems create electromagnetic fields, and yet we all feel better around waterfalls, trees, garden plants, animals, etc., unless we have some allergy. Every *human-made* device that uses electricity or generates electrical signals creates EMF. However, there's a difference between natural EMF and human-made EMF.

The human-made EMF creates unseen chaos in the environment. Every power line, light, computer, cell phone, motor, building, vehicle, appliance, power tool, TV, music system, video game, radio waves and wireless technology, create chaos and stress. Their effects are opposite of LOVE and coherence. Nearly every living organism on Earth lives in this toxic chaos that causes incoherent consciousness and societal stress. Why is that?

Plague of Measurement Uncertainty

All technology using electricity is based on the approximation of the measurement of frequency (time) and amplitude (power). The functioning of a system is complicated through the interactions between each signal and field. When measurements are based on approximation, which is the common standard for measuring time, each measurement is incongruent with every other measurement. This causes measurement uncertainty. Consequently, it becomes a big costly mess of *phase noise.*

Measurement uncertainty causes *discrepancies* of time and power, or frequency and amplitude, in systems. These discrepancies of frequency and amplitude over time and distance amplify and compound noise. This causes havoc in technological and biological systems. That includes electrical power and information systems which affect every individual, society, and environment.

Electromagnetic noise is also an *irritant* and an unnecessary source of stress and chaos. It compromises safety, health, efficiency, and productivity in human beings, animals and plants. Collectively, this wastes huge amounts of energy, life, and financial resources, and is far less sustainable. This affects everyone to some degree, even if you don't know it. So, the primary problem is not electromagnetic fields themselves, but chaos within these fields that causes *noise* and therefore *stress*.

Coherent Electromagnetic Fields

Different technologies attempt to address the problems of EMF chaos and noise. Each of these technologies claims to mitigate the EMF problem in a unique way. Many technologies simply filter out higher frequencies that degrade easily and cause more noise. However, those approaches lose important signal data.

Traditionally, there are three ways to improve an electronic circuit. You can improve the design, improve the materials, and improve the ambient conditions in which the circuit operates. New discoveries in monatomic materials, such as graphene, improve the *materials* part of that equation.

Rise of *COCUUN*

A fourth way of improving an electrical circuit has created startling results. One technology reduces such noise by increasing accuracy, precision, stability, and integrity of signals so that data and power are not lost through incoherence. I call it *COCUUN*™, which is short for *COherent CUrrent UNion*™. The word "union" refers to devices, connectors, or some other interface that converts incoherent current (electricity, fluid, gas, or plasma) into coherent current – as kind of a "metamorphosis", or a "*COCUUN Effect*". Trademarked logos:

The inventor of *COCUUN* devices claims that they harmonize artificial EMFs with humans and other living systems. Objective measurement, coupled with subjective experience, shows vital benefits of this technology for a wide spectrum of applications. That includes reducing chaos or signal loss in power lines, motors, microchips, audio and video signals, AI, and much more.

Increased Brain Wave Coherence

Our brain generates electrical activity in waves. An electroencephalograph (EEG) measures relative coherence of brain waves. *COCUUN* tech increases EEG-measured brain wave coherence. Research has correlated coherent brain waves with better school grades, increased creativity, improved task performance, faster reaction time, higher moral reasoning, and increased IQ.

Users of *COCUUN* technology who work around computers reported:

- More clarity of thought, focus, creativity, and serenity
- More energy and stamina, both mental and physical
- Relief from computer-related discomfort
- The ability to work more productively and effectively
- More harmonious relations with others
- Greater success in task performance
- Better sleep

Applications to Electronic Media

COCUUN tech reduces noise in video and audio signals. Do you listen to music, play video games, or watch movies, television, or Internet videos? Noise, distortion, and signal instability, are inherent in all of these media. Signal noise irritates living systems.

The entertainment industry has unknowingly contributed to incoherence in collective consciousness. That causes degradation of human life. This noise is inherent in all electronic media productions. It's a major source of the virus of incoherence.

COCUUN technology reduces this noise. The entertainment industry's *Sustainable Entertainment Alliance* could raise the zone of society, inoculate audiences from the virus of incoherence, and basically "save the world" in real life. Those achievements would be side-benefits of using *COCUUN's* coherent current technology to upgrade the performance of their electronic equipment.

One *COCUUN* device improved a radio station's signal so much that it reached a major city that was previously out of range. Some listeners said it was like live studio music. A radio signal that goes farther with more clarity means a bigger audience and more *satisfied* listeners. That means more advertising revenue for radio or TV stations, which can then also raise the zone of society.

COCUUN turns sources of EMF chaos and stress into sources of LOVE and coherence. Increased coherence and reduced stress improve health and increase loving and kind behavior. That also improves human performance in all environments. The application of *COCUUN* for creating coherent electromagnetic fields is a major strategy for creating macroscopic coherence, even on a global scale. Let's consider two more strategies.

Coherent Fuel Plasma

Clean air is one of the vital resources for human brains to work the best and have high levels of coherence. Numerous scientific studies have found air pollution to be a causal factor for cognitive decline, dementia, mental health disorders, and negative effects on brain development in children.

Another *COCUUN* coherence technology offers a highly affordable solution to air pollution. It converts incoherent fuel into coherent fuel plasma, making fuel combustion more efficient and complete. Third party validation testing on a diesel truck showed:

- 15% increase in fuel efficiency
- 40% increase in torque
- 40% decrease in hydrocarbons
- 66% decrease in particulate matter 2.5

Vehicles as Air Purifiers

In environments with extremely bad air pollution, such as in China and India, vehicles using *COCUUN's* coherent fuel plasma technology could potentially produce exhaust that is cleaner than the air they take in. That would convert vehicles, even race cars and semi-trucks, into mobile *air purifiers* instead of *air polluters*. Not even *all-electric* vehicles are able to achieve that! These compact devices are for any size and type of engine that use air and a hydrocarbon fuel for combustion.

Coherent Water

Water is essential for life on Earth. Everyone knows that fact. Many indigenous populations revere water as sacred and even as a living entity with memory, healing powers, self-purification, and more. However, the mass industrialization of human civilization has desecrated water to dangerous levels.

In response to that global crisis, different methods have sprung up to restore water to its more natural and health-promoting state. Many of these methods cite the pioneering work of Austrian German, Viktor Schauberger and his exploration of "living" water. Grander Water Revitalization is the first such technology that I learned about. A more recent development is "coherent water" by a company named Analemma.

The scientific research at Analemma supports their claims that coherent water has beneficial effects for cellular energy, gut microbiome, plant and soil health, and brainwave coherence. The latter area of research is especially fascinating and relevant to this part of our tour. We've already spent a good amount of time on the topic of brainwave coherence because that is the foundation of our ability to raise the zone of collective consciousness. Again, most of humanity experiences common sources of electromagnetic fields, air, and water. So, now we'll envision ways to apply these coherence-creating technologies on global scale.

Creating a Global Coherence Grid

Applying *COCUUN* technology to power grids could decrease societal stress caused by incoherent electromagnetic fields. A "Global Coherence Grid™" project would measure if *COCUUN* devices could increase sociological coherence for the population an entire major city. If such a hypothesis is proven, the goal would be to convert networks of power grids, globally, from creating stress and chaos to reducing stress and chaos, while increasing measurable levels of coherence. Increasing EMF coherence in power grids would also enhance the performance and cooling of data centers and quantum computing, where coherence is vital.

Testing this hypothesis would require analyzing scientific data to confirm if *COCUUN* technology appeared to increase quality of life indicators. It would include statistics on health, crime, violence to humans and animals, accidents, suicides, and other measurable social indicators. Even air quality would be monitored to validate the novel hypothesis that increasing coherence in power grids can reduce air pollution. That's the primary topic of Chapter 13.

Acute and chronic stress kills brain cells. It can damage DNA, hormones, brain functions and all aspects of health. That includes emotions, memory, learning, and the prefrontal cortex ("CEO" of the brain) which filters destructive thoughts to prevent violence.

Previous studies on reducing stress and increasing coherence society-wide have shown reductions in all categories of violence, crime rates, child abuse, substance abuse and more. Successfully demonstrating this "coherence grid" would offer hope for a global improvement in quality of life for populations negatively affected by electromagnetic fields, which is most of humanity.

Confirming the Global Coherence Grid hypothesis could transform unintended stress-inducing hazards of electricity into perpetual global generators of greater LOVE and coherence. That, in turn, would increase the capacity to LOVE even more, all in a mutually reinforcing upward spiral. It would also raise the zone to elevate government leaders to serve the public better.

1,000 Cities of LOVE and Coherence

Quality of life indicators would likely improve throughout the proposed one-year project. Therefore, a Global Coherence Grid created in our world's 1,000 largest cities could benefit most of Earth's human population which lives in urban areas.

COCUUN tech could be deployed within a few months or less. It would not take years. Cost factors would include supply and demand, competition, urgency, and further development of these *COCUUN* devices.

Fifty years ago, computers filled entire rooms and cost hundreds of thousands of dollars. Now you can get even more computing power in a "smart" watch for a few hundred dollars.

"Confirming the Global Coherence Grid hypothesis could transform unintended stress-inducing hazards of human-made electricity into perpetual global generators of greater love and coherence."

Music for Healing and Balance in Nature

Our environment experiences stress too. Scientists talk about stress on crops, forests, and animals. Ancient "Solfeggio" frequencies were rediscovered by the late Dr. Joseph Puleo, and by Dr. Leonard Horowitz. Frequency generators, musical instruments, and chanting, can create these frequencies. They allegedly have uplifting and healing benefits, including DNA repair, which is claimed to result from the 528 Hz frequency.

Every culture has music traditions to create healing, harmony, and balance. According to Maharishi Ayurveda Products International, playing or performing a "Rain Melody" of Gandharva Veda Music helps to bring rain. It achieves this by replicating the frequencies present in nature when it's raining. Gandharva music uses sound, melody, and rhythm to restore balance and harmony in mind, body, behavior, and environment.

In the mid-1990s, I organized tours for Maharishi Gandharva Veda musicians. I heard stories from different locations about rain starting during concerts, even when rain was not predicted. Using a historically successful approach to increase the chance of rain could lessen death and destruction from wildfires.

TV and radio stations could broadcast the Rain Melody for at least part of the day in areas of drought and forest fires. It would reduce societal stress as well. It doesn't matter if you're listening to it, but it's more powerful if you do, particularly at a live concert.

In addition to EMF chaos, all negative thoughts, feelings, and actions also add toxic stress to society and our environment. These stress-reducing technologies offer profound leverage for anyone to improve their inner and outer environments.

Creating a LOVE-Based Economy

All societies have an economy and a government. Today, both fall far short of their potential to serve humankind. However, it's not enough to change our system of economy or government to become a solution-based civilization. The word "economy" has too limited of a definition anyway. It includes money, investments, stock market, Gross Domestic Product, exports, imports, trade, jobs, and other such things. All of these are based in scarcity.

Exploring the history of money and economics is a daunting task. Fortunately, some brilliant and insightful people have completed that task. Charles Eisenstein's book: *Sacred Economics: Money, Gift, And Society In The Age Of Transition* is one of the best examples. He explains where we've been and where we need to go with money and economic systems.

In a kind and compassionate way, Eisenstein has revealed the damage from our lust for money and materialism. He describes a well-lit roadmap of wiser stewardship of our lives and planet. He envisions a LOVE-based sacred economy that reunites humanity in greater connectedness, which money alone cannot provide. His insights about complex systems are deep and comprehensive.

His chapters include topics such as The Illusion of Scarcity, Negative-Interest Economics, The Social Dividend, and Local and Complementary Currencies. However, Eisenstein is not calling for radical disruptions to existing economic systems and currencies. Instead, he offers a practical way to quickly evolve a more livable and humane civilization of abundance for all.

Our human resource of thoughts, words, emotions, actions, physical health, and creativity are the most important parts of our life, purpose, and economy. The old economy was fear-based and depended on control. The new economy is LOVE-based and draws on true power. Fear controls, and LOVE inspires.

In the old paradigm, the measure of wealth was how much you rose above others by acquiring money and property. In our emerging paradigm, the measure of wealth is how much you can uplift humanity along with yourself.

Truth, coherence and LOVE are currencies of this new economic paradigm. They're unlimited resources. What's driving this transformation in our economy? It is the nature of life, which is to expand. That translates into the desire for more happiness.

You might ask, "How is that different from people hoarding money, exploiting others, and making resources scarce?" Well, have you ever gone down a road that led to a dead end at a wall or a cliff? Did you just keep driving into the wall or off the cliff simply because you had momentum in that direction?

We have gone down that old paradigm road for too long. Now it's time to back up, turn around, and find another way forward. Our desire for greater happiness is still the driving force. But now we must drive in a more economically and sustainable direction to create a new journey and destination that helps *everyone.*

It's like the Butterfly Effect and putting ice crystals in a pond. What do you have when you combine these two principles? Think carefully about your answer. If you say, "frozen dead butterflies," you'll need to start our tour all over. What you have is an opportunity to create a major global transformation in the economy – quickly and efficiently!

Do you also remember our discussion about *selfish, selfless, and selfness?* The same principle applies here as well. When people spend money or invest within the sphere of their self, it can appear either greedy or generous. If their sense of self is small, then it seems greedy to others who have a larger sphere of identity. If their self-identity expands with their increasing connection to the universe, then they become more philanthropic.

We need a solution-based civilization as the foundation of all wealth, not merely a different economic system!

How to Manage Your Greatest Natural Wealth

Are you familiar with Lego blocks? Imagine that every thought, word, or action uses one Lego block, and you had limited blocks? Pretend that these blocks build your life and wealth.

Now, imagine that every time one of your thoughts, words, or actions is negative, hateful, disrespectful, greedy, or dishonest, the block floats away. It joins other blocks that build problems in your life and your community. These problems, in turn, create businesses such as the weapons industry, disease care industry, pharmaceutical industry, prison industry, fossil fuel industry, corrupt financial and insurance industries, and so on. In short, anything you do that's opposite of creating truth, coherence, and LOVE will feed, fuel, and fund these life-damaging industries and other structures in a problem-based society.

Now, picture every thought, word, emotion, and action that expresses various aspects of LOVE and coherence. These blocks build solutions. These solutions include prosperity; cleaner air, soil and water; and harmonious communities. These results, in turn, build life-supporting businesses, healthy food, schools, clean energy technologies, natural health care, and provide basic living necessities for everyone. LOVE and coherence raise the zone of the economy in an upward spiral of reinforcing frequencies. You could call it "bubble up" economics, instead of "trickle down" economics, which mostly benefitted the wealthiest population.

"Bubble up" economics can be understood with a simple, yet profound, analogy. Everything we do that creates expanding spheres of LOVE and coherence, gives buoyancy to innovation, inspiration, and expanded thinking. These are the essential building blocks of an economy. These special spheres, or bubbles, of LOVE and coherence are powerful. They remind me of the 1970s idea to "raise the Titanic" by filling it with ping pong balls. That's similar to how we can raise the zone of global economies.

Wouldn't you rather build a better quality of life for yourself, your family, community, and humanity? It all depends on choices you make, and the positive or negative quality of your thoughts, words, and actions.

Light Up a Global Transformation

Are you getting the big picture of possibilities here? You have the power to initiate extraordinary solutions to big and pervasive problems. It takes only the right combination of readily available solutions and commitment to create global LOVE and coherence. You already have enough resources and power inside you to help achieve a global transformation.

There's an old saying, "The world is run by those who show up." How about a new saying, "The world is transformed by those who raise the zone"? Are you ready for that? You don't need to wait for anyone else, certainly not the government.

However, when you create more coherence and LOVE in collective consciousness, then all governments will create better policies to ensure more coherence, prosperity, and well-being for all citizens. *You* are the one you have been waiting for. What's your push-button solutions for global change? Share them!

This chapter proposes a "do-it-yourself" approach to creating global LOVE and coherence. It's affordable, practical, and based on breakthroughs in science and technology, as well as human choices. The miraculous power of commitment works the same for big goals as it does for small goals. So, think big!

Among all of the solutions shared in this book, LOVE is truly the simplest and highest technology for creating harmony. LOVE is yours to access at any time. You don't need permission, special passwords, or a secret handshake. It can't be stolen, hacked, or even diminished by anyone else. Only *you* can suppress the LOVE you have. That is important enough to repeat. **Only *you* can suppress the LOVE you have!**

Have you been suppressing LOVE, or keeping it under wraps for a special moment? Well, as your tour guide, let me say that moment is now! Rip the wrapping off your heart. Give your gift of infinite LOVE and radiate your light of LOVE, now! There, you did it! Can you just feel the LOVE in this room?

Summary

Negative thoughts, actions, and behavior contribute to acute and chronic stress in our society. Other contributors are chaos in electromagnetic fields, environmental pollution, and the lack of truth, coherence and LOVE. Stress creates resistance to a natural and spontaneous flow of LOVE and our state of coherence. It also creates dysfunction in mind, body, and behavior. All of that impedes the economy and prosperity of our lives and society.

Truth, LOVE, and coherence are the most effective antidotes to stress and the virus of incoherence. Various methods can create coherence on the larger scales of societies, nations, and our world. Each one creates a rapid upward spiral of benefits to humanity. All of that lays the foundation for a new paradigm of enlightened, solution-based economy. Like the Butterfly Effect, the LOVE you express can stir positivity, and spread it to support humanity's rise in consciousness, peace, harmony, and prosperity.

Next, we'll explore a global pollution solution using mostly the technology and devices of *COCUUN* for electrical coherence. This was also the basis for the Global Coherence Grid. Shall we do an amazing "class project" to end our *RAISE the ZONE* tour?

13

Push-Button Air Pollution Solution

"Give me a place to stand,
and a lever long enough,
and I shall move the world."
- Archimedes

Lift Air Quality with an Archimedes "Lever"

Is it possible to reduce global air pollution with the push of a button? Actually, yes, although it could also be a lever, or switch, or some other way to "turn on" coherence at a macroscopic scale. *COCUUN's* amazing new coherent EMF "lever" can "move the world" above deadly levels of air pollution easily and quickly. The *COCUUN* coherent fuel plasma technology would require wide deployment of the devices as standard equipment on internal combustion engines. Each of these technologies are able to reduce air pollution more affordably than other approaches.

The *COCUUN* device, which increases coherence in electromagnetic fields, also reduces air pollution. Air quality levels are precisely monitored in real time, many times a day, all over the world. While climate change and air pollution are related, climate change is not measured as easily, frequently, and precisely, as air pollution is measured.

World's Population Chokes on Air Pollution

How bad is global air pollution? It's bad enough to make it the second leading cause of death for children under age five. The World Health Organization (WHO) has declared that:

> "92% of the world's population lives in places where air quality levels exceed WHO limits." It has also stated that, "Some 3 million deaths a year are linked to exposure to outdoor air pollution. Indoor air pollution can be just as deadly."

Allowing Children to Breathe

Do you agree that allowing children to breathe cleaner air is urgent? Air pollution can also damage fetuses, developing brains, and other organs in children. On March 6, 2017, WHO reported that 1.7 million children die every year from environmental pollution. One third of those are from respiratory disorders, much of which is due to indoor and outdoor air pollution.

The *State of Global Air/2024 - A Special Report on Global Exposure to Air Pollution and Its Health Impacts, with a Focus on Children's Health,* issued similarly shocking assessments. It was a collaboration between the Health Effects Institute and the Institute for Health Metrics, and Evaluation's Global Burden Disease project. The effort was in partnership with UNICEF.

More children under age five die each year from air pollution, than the total number of American's who have died in combat since 1775. This global problem isn't due to a lack of affordable and effective solutions or funding – it's mostly political.

Citywide Electrostatic Trap of Air Pollution

The inventor of the *COCUUN* coherent EMF device had a novel hypothesis. He believed that chaos, or noise, in electromagnetic fields from power lines and building wiring created a "spider's web" matrix of electrostatic traps that held air pollution close to the ground. He set up a three-month pilot project in 1990, from August to October in Los Angeles, California. At that time, 11 million people lived within the 13,600-square mile Los Angeles area of the South Coast Basin. **The American Lung Association ranked Los Angeles as #1 in ozone pollution for 25 of the last 26 years.**

He went to Anaheim and plugged his device into a common electrical outlet for L.A.'s power grid. Then he monitored publicly available data on air quality. Within about two weeks, the poor air quality improved by an average of 43% in the Los Angeles metropolitan area. Daily pollution levels fell below predicted levels, and also below federal limits. They remained low until the end of the three-month study.

After the device was removed from L.A.'s power grid, levels of pollution returned to normal levels for that time of year. Local air quality officials were mystified. The senior meteorologist with the South Coast Air Management District, quoted in the Los Angeles Times, said:

"This is mind-boggling. When you compare it with the past, it's like night and day. There's really been a dramatic improvement."

The project's results suggest that creating more coherence in power grids could significantly reduce air pollution. The inventor also hypothesized that EMF chaos in power grids, and everything connected to them, creates high-pressure zones in urban areas that disrupt natural patterns of rain. His initial success warranted more research, but the inventor explored other novel applications for his invention instead.

The L.A. air quality test in 1990 didn't require any burden on taxpayers and ratepayers. The anti-pollution measures of today would not be as necessary if technologies that generate coherence in power grids had been in use these last 30+ years. Feeding such coherence into the power grid can also be achieved by multiple smaller devices. It's good to have some redundance anyway.

How "Greased Lightning" Could Clean Our Air

The scientific explanation of *COCUUN* coherent electricity is forthcoming. The "secret sauce" of the invention is proprietary. So, just think of that common phrase, "faster than greased lightning." Lightning is electricity. When electricity flows through power lines, wiring, and circuits, there's resistance. It's due to friction from physical materials and measurement uncertainty in electrical systems. That's true even for electricity made from wind, solar, or other clean energy sources. Friction causes heat and energy loss. Imagine if you could "grease", or lubricate power lines internally, similar to lubricating an engine with oil to keep it from over-heating due to friction. Cooler and smoother flowing electricity would create more power with less effort, wasted energy, and pollution.

Shortcut to Reduce Greenhouse Gases

Is there any *other* evidence that *COCUUN* tech could offer an affordable way to decrease carbon emissions and other greenhouse gases? Yes! Electrical inefficiency causes energy loss and excess heat. In 2010, a paper in the *Renewable and Sustainable Energy Review* stated that, "Statistics indicate that electric motors are generally responsible for about 2/3 of industrial power consumption in each nation, or about 40% of overall power consumption."

The paper also stated, "Previous studies have reported that implementing a few options with little or no cost in the industrial sector could reduce 10–30% of GHG emissions." It added that, "In many industrialized countries, more than 70% of the total produced energy is used by electric motors."

Industrial motors heat up due to friction, resulting in energy loss. Increased electrical efficiency reduces waste heat and lost energy, which results in less pollution and fuel use. That means less GHG emissions. Attaching *COCUUN* devices to electric motors has already demonstrated the ability to reduce their running temperature by about four degrees centigrade (7.2° Fahrenheit.)

The combined effects of increased electrical efficiency from *COCUUN* tech in power generation, distribution, and consumption could therefore achieve far greater reductions in global GHG emissions than the 10-30% estimated for electric motors alone. It would also be "with little or no cost".

Time is of the essence when it comes to life-threatening risks of polluted air and catastrophic climate change. Don't buy into claims that responsibly dealing with global air pollution, or even climate change problems, will take decades or centuries. So, how do we go forward with such breakthrough solutions?

Global Carbon Reduction Grid

Technologies such as *COCUUN* can truly "move the world", as Archimedes claimed. Think of our existing power grid infrastructure as a fulcrum, and *COCUUN* tech as a lever. You merely plug the device into an electrical outlet of a power grid, or maybe at the power generation plant itself.

You power up a *COCUUN* device with the push of a button or flip of a switch. The inherently incoherent current starts to become more coherent. Chaos and noise are "lifted" from power grids' electromagnetic fields. Numbers projected in the following chart for a proposed global project are for only one type of device, such as *COCUUN* tech for creating coherent electromagnetic fields.

- **Only 1 device was used to "lift" (reduce) air pollution an average of over 40%** from 11 million L.A. area residents during the three-month project in 1990.
- **Only 10 devices, in the 10 most populated cities in the world, could benefit over 250 million people.** Within a few months, they could have better air quality, improved grid efficiency and other quality of life benefits.
- **Only 1000 devices, in the 1000 most populated cities in the world, would benefit over 2 billion people** within six months.

Proposed					
Number of	**1000**				**2 billion**
Metro Areas	**100**			**1 billion**	
with *COCUUN*	**10**		**250 million**		
Power Grids	**1**	**15 million**			
		Approximate number of people benefitted			

On May 1, 1893, President Grover Cleveland flipped a switch, or pushed a button, at the Chicago World's Fair. Instantly, nearly 100,000 lamps lit up in that epic demonstration of Nikola Tesla's A/C motor invention which became a global game-changer.

Global Outdoor Air Purifier

Envision an ambitious project to quickly deploy *COCUUN* technology in the world's 1,000 largest urban areas. According to the *DEMOGRAPHIA WORLD URBAN AREAS*, that would account for about one-half of the world's urban population. That equals approximately two billion people as of 2024.

COCUUN devices are a viable and highly affordable solution to global air pollution. This would decrease COPD, asthma, and other respiratory problems. *COCUUN* tech could reduce the loss of school days for asthmatic children, and decrease related lost workdays for working parents who care for such children.

Using coherence-generating technology to "tune up" power grids turns them into outdoor air purifiers. Benefits to air quality would be measurable within weeks. Such devices are scalable for any population size. Within a short time, *COCUUN* devices could upgrade all 50,000 power plants in operation, and their respective grids, worldwide. Among those, 7,000 are in the U.S.

The Los Angeles Air Quality study used only one device about the size of carry-on luggage. How could it create coherence so quickly throughout the power grid for an entire metropolitan area? It's just the exponential power of constructive interference.

Coherent Exponential Growth

In laser light, the square root of photons acting in-phase, i.e., coherently, instantly stimulates all other photons into a macroscopic wave of coherent light. Consider two examples of rapid growth. There's a well-known classroom exercise where a teacher asks, "What would you rather have, $1,000,000, or one penny that doubles in value each day for thirty days, and then you get the cumulative amount at the end?" Students often say $1,000,000 sounds best to them. Then their teacher explains that if you doubled a penny's value each day for thirty days, it would be $5,368,709.12 at the end of the 30 days! Big lesson learned!

During a human embryo's development in the womb, one cell continues to divide to form an entire complex human body in about nine months. The number of cells at the end is staggering. *Smithsonian Magazine* stated that the adult human body has 32.7 trillion cells! It added that such an estimate is tricky because of different ways of calculating, which could range from 15–70 trillion cells. Even a newborn baby could have at least 26 billion cells.

That's the transformative power of LOVE and coherence. In many cases throughout history, just one person set things in motion that impacted and transformed our whole world.

Okay, back to our explanation of rapid exponential growth. Does the penny example help explain how coherence could grow in a power grid? Doubling coherence each day would reach the following number of people, starting with just one person:

- **Day 1 = coherence affects one person. Day 2 = two people. Day 3 = four people, etc.**
- **Day 23: 4,194,304 people (more than the population within L.A. city limits in 2025)**
- **Day 25: 16,777,216 people (L.A. region = about 13,000,000 people in 2025)**
- **Day 26: 33,554,432 people (The estimated population of California is over 39,500,000 in 2025)**

That roughly equals the speed of improved air quality in the L.A. pilot study, using just one *COCUUN* device. Here's another example. Would the Fibonacci sequence of numbers describe the speed of improvement better? [0, 1, 1, 2, 3, 5, 8, 13, 21, 34, ...etc.]

- **Day 34: 5,702,887 people (more than the population within L.A. city limits in 2025)**
- **Day 36: 14,930,352 people (L.A. region - about 13,000,000 in 2025)**
- **Day 38 = 39,148,169 people (The estimated population of California is over 39,500,000 in 2025)**

Manufacturing and installing enough coherence devices for power grids worldwide would be relatively simple, quick, and even profitable due to financial benefits of reducing carbon footprints. It wouldn't require acts by the United Nations, heads of state, legislative bodies, or non-profit organizations. Even burdensome emissions regulations would not be required.

Improving air quality, and overall quality of life, doesn't even require a majority vote. Anthropologist Margaret Mead claimed that the only thing that ever changed the world was small groups of thoughtful and committed citizens. Even a single visionary philanthropist could create a Global Carbon Reduction Grid.

Global Clean Air "Earthshot"

Reducing air pollution by 25% in the 1,000 largest metro areas within 3-5 years is very feasible, almost with the push of a button. Creating a Global Outdoor Air Purifier is my moonshot, which is actually an *Earthshot*. It could be proposed for the Earthshot Prize that "finds, supports and celebrates those who turn bold ideas into solutions for our planet." [See: earthshotprize.org]

This "Earthshot" would also produce other personal and societal benefits in addition to reducing air pollution! People would say, "Wow, we did that? What else can we do? Let's think bigger and aim higher!" Have you decided on your moonshot?

Getting Big Drivers in the Race

Why would big *drivers* of the economy want to join a global race to reduce air pollution? Think of motorsports organizations, auto manufacturers, transportation fleets, and utility companies. The simple answer is more *money!* They can spend less on their emissions reductions. However, they can also achieve their goals for better air quality at a much faster *speed*!

Motorsports: Formula One (F1), Indy 500, NASCAR, and especially Le Mans 24 Hour Race, are all lowering their carbon footprints. They're bringing in race cars that are hybrid gas/electric, all electric, and hydrogen-fueled. They're also reducing their emissions and carbon footprints across their entire operations.

However, they're facing strong resistance from fans and race car drivers. Potential lithium battery fires create another issue. Altogether, that's making it a rough road. *COCUUN* technology can greatly reduce exhaust emissions. That would help motorsports achieve goals for emissions and profits, while pleasing fans and race car drivers.

Automotive industry: auto makers look to motorsports for new ideas in production vehicles. At least 80% of all vehicles will have internal combustion engines for many years ahead. And yet, manufacturers are struggling to reduce emissions. The transition to hybrids and all-electric vehicles has been slow and fraught with problems. It's also disrupted by uncertain markets and supply chains, not to mention volatile government policies.

Many electric vehicles purchased in the U.S. have relied on financial incentives and government grants. Those are being reduced or eliminated. Emissions goals are being threatened. The *COCUUN* coherent fuel plasma devices can help automakers achieve emissions goals while increasing profits.

Transportation - The owners of motorsports organizations, such as John Malone of Liberty Media and Robert Penske of Penske Corporation, are also lowering emissions throughout their other businesses. Electric semi-trucks are problematic due to lack of charging infrastructure, and the massive weight of batteries which reduce loads that can be transported. *COCUUN* coherent fuel plasma technologies can reduce emissions for trucking companies. Saving an average 15% on fuel costs means tens of millions of dollars in new profits annually for large companies.

Pepsi and Coca Cola have lucrative contracts with schools across the U.S. Even a tiny percentage of profits from saving 15% on fuel costs, could easily pay for dramatically reducing harmful emissions of school buses and public transit buses. School districts and municipalities always struggle with budgets. Increased fuels savings from *COCUUN* devices (<$2000/each), and possible carbon credit income, could help them achieve their air quality goals more economically. That's far more sensible than schools spending $400,000 for each new electric school bus. Public transit EV buses cost even a lot more than school buses.

COCUUN vehicle devices could help achieve nearly the same reduction in life-cycle greenhouse gas emissions as EV buses. The simple fact is that goals for reducing harmful vehicle emissions can be realized without any more government grants, subsidies, or even electric vehicles added to the global inventory.

Electric utility companies: Demand for electricity surges daily for data centers, AI, and cryptocurrency. It's also rising from the increased electricity use in transportation, home heating, cooking, and hot water. Utility companies have rising costs from lawsuits and emissions regulations. These costs get passed on to ratepayers, investors, and the economy. Adding *COCCUN* tech to power generation, distribution, and consumption, is the most cost-effective solution to this crisis. It would reduce energy costs, while increasing profit, safety, and performance.

All of these scenarios benefit the private sector, consumers, and environment. That makes it highly attractive and feasible for the private sector "big drivers" of the economy to support this "push button" Global Carbon Reduction Grid project.

New Science – Not Merely New Technology

The epic discovery of transforming incoherent currents of electricity, fuel plasma, and water, into coherent currents, is vital for every living system on Earth. Applied coherence technologies will be essential programs of study for all levels of education. Each one warrants extensive research across every field of human endeavor at every scale. Yet, they already provide the means to *raise the zone,* clean up our M.E.S.S., and light up a luminous age of humanity. They are part of a *new science,* not merely a *new technology!*

We are the "Great Leader"

Many people are still waiting for some "great leader" to fix what's wrong in our lives and our world. Who is that great leader? "We, the People" are that leader. You are the leader of your life. Hopi Elders said that **"We are the Ones we've been waiting for."** We don't need to wait for anyone or everyone. Never has everyone taken a step forward all at once. Some led the way. Others took their own steps later. The more LOVE and coherence we create, the better rulers we become. Then politicians will follow *us* and our government will be uplifted as we raise the zone.

Humanity is already the greatest power grid on Earth. That's not just a metaphor. The HeartMath Institute has discovered that each of us links electro-magnetically to others. HeartMath's Global Coherence Initiative, and the Institute of Noetic Sciences' Global Consciousness Project, are validating this global link. Physicist Nassim Haramein has mathematically described our connection to the entire universe. Nature's infinity is also *our* infinite nature.

Tragically, our connectedness to each other and the cosmos has become horribly dysfunctional and in dire need of repair and upgrades. It's time to update our software with coherence and LOVE. We've been running outdated programs that can't get us to a future that we desire and deserve. Creating an enlightened civilization and responsible government is in our hands.

The Giving Xperiment inspires compassion, gratitude, courage, humor, and forgiveness, just to name a few values. Each one of these acts of LOVE is available in each moment. Sometimes it might seem that LOVE is almost nonexistent. Has our intelligence outgrown our wisdom? Have we lost sight of truth and LOVE? They have not gone anywhere.

Maybe we can't bear to see so much pain and suffering in our world, so we close our eyes to these things. Wisdom, truth, and LOVE are the elixirs that will heal a sick and suffering humanity. It's time to expand our vision. It's vital for us to see the new and exciting paradigm shift that's underway. Now is the time for your light to illuminate your world and raise the zone for humanity!

Passing the Baton – The Race is On!

You have all that you need to transform your world into what you would like it to be. This tour is almost over. Now we celebrate! You have your own personal guide. That guide is YOU! You're certified. Your tour is just beginning, accelerating, or expanding in scope and purpose. I'll be cheering for your success!

Keep expanding your thinking and radiating truth, coherence, and LOVE at every opportunity. You don't need any moonshot project to get started. Even small actions can get you to "escape velocity" and overcome the gravity of fear or doubt. Every day you defy gravity when you sit up or stand up, lift an arm, put food in your mouth, take a step forward, or help someone else to rise up. It's second nature to you. Every moment that you overcome the gravity of fear and doubt, with the light of LOVE and coherence, you raise *your* zone, and increase momentum to achieve your dreams – big and small.

The People's "First 100 Days"

The first 100 days in office of a newly elected President of the United States of America is a big deal. At least the news media and other observers think so. What would you do as President?

Do you ever wonder how much "We, the People" could achieve in *our* first 100 days* of being in charge? Actually, we *are* in charge. Do you recall U.S. President Lincoln talking about our "government of the people, by the people, [and] for the people"? [*See Appendix A for ideas.]

Call to Action and a Word of Caution

RAISE the ZONE is a call to action. It's a command to fully express the highest and best of yourself and humanity. Even small efforts to increase truth, coherence, light and LOVE could create magnificent benefits that ripple across our world to achieve a lasting era of Earth's Luminous Age – kind of a *Heaven on Earth*. You might also find that "Heaven on Earth is a tough town!". That means it could be more wonderful than you can imagine, but keeping it that way requires *vigilance* and *commitment*.

Countdown to Blastoff

Are you ready to launch your moonshot or *Earthshot*? Are you ready for one of the most extraordinary adventures of your lifetime? If you are, all systems are "Go"! Ten, nine, eight...well, you know the rest. Stay curious. Seek truth. Take action. Think big. Aim high. Be luminescent. Commit and persist!

> "Whatever you can do or dream, you can begin it.
> Boldness has genius, power and magic in it!" - Goethe

Please get your "Tour Souvenir" on the next page. **Thank you for joining this tour. Now it's up to you!**

Light Up and Shine On!

Illuminate your solution-based world!

RAISE the ZONE!

RAISE the ZONE
Tour Souvenir

528 Hz LOVE Frequency

Thank you for joining me on this *RAISE the ZONE* tour. You made it to the end after enduring my tour jokes and all. I am truly impressed! Sharing this journey with you has been a wonderful experience. I hope you enjoyed it too.

In parting, I leave you with a special LOVE tour souvenir. It is your awareness of the 528 Hz frequency. Researchers, musicians, and historians have discovered that unconditional LOVE resonates at 528 Hz frequency.

This LOVE frequency originates in nature. According to some researchers, it is also a "miracle" note in the original Solfeggio musical scale. 528 Hz frequency is associated with DNA repair and both personal and environmental health. You can listen to it on Internet videos and enjoy the benefits.

Remember that truth, coherence, and LOVE are the most important keys to your health, happiness, and wealth. Upward and onward to your light-filled life and solution-based world!

Sincerely,

Jeff Peckman

Acknowledgments

This book is like one of those photo mosaics that is made from thousands of photos that look like someone's face. Thousands of life experiences, and influences from people I love and respect, helped create the magical mosaic of my first book, *POWER UP – How to Create a Solution-Based World in 100 Days.*

RAISE the ZONE is a major revision of *POWER UP*. It contains content that might be viewed as controversial, and is not known to those who contributed to *POWER UP*. Unfortunately, we live in volatile environment today. There's too much possibility of attacks on people with controversial viewpoints.

Therefore, I have chosen to not acknowledge the names of those who has been part of the "magical mosaic" of my life that led to RAISE the ZONE. You know who you are and that I'm deeply grateful for your support. Thank you!

Glossary

The most relevant definitions and synonyms for some key words used in ***RAISE the ZONE*** are below. All are from a Google search unless otherwise noted.

Coherence/Coherent: The quality of being logical and consistent... a unified whole, (synonyms: rational, clear, lucid, and articulate. In physics: pertaining to waves having a constant phase relationship.

Harmony: Accord, peace, cooperation, unity, oneness, balance

In-phase: Of, or relating to, electrical signals that are in phase. Being or happening in synchrony or harmony. [Related to "constant phase relationship" in 'coherent']

Light Up: Illuminate, become lively with joy, to shine, to make bright, to radiate light.

Love: An intense feeling of deep affection, (Note: synonyms of non-romantic LOVE: Friendliness, kindness, charity, goodwill, sympathy, altruism, unselfishness, philanthropy, benevolence, humanity)

Power-up: The action of switching on an electrical device, especially a computer.

Raise the zone: Increasing coherence in individual consciousness and collective consciousness of society. [Author's definition]
Synchrony: Simultaneous action, development, or occurrence. The state of operating, or developing, according to the same time scale as something else, (synonyms: harmony, coordination, integration)

Appendix A
The People's "First 100 Days"

What can "We, the People", do in 100 days to "raise the zone" and create a solution-based world?

"We, the People" have all of the authority and resources we need to raise the zone and create a solution-based world. We just need to light up as enlightened solutionists! Below is a short list of things you can do and see what results.

1. Learn about COCUUN™ coherent current technologies
https://www.COCUUN.world

2. Learn About and Express Unlimited LOVE

Participate in the Giving Xperiment (Gx) Surveys and Set up https://oneworldheartproject.com/

Read *Better Than Cheese* featuring Amazing Mouse
www.amazingmousebooks.com

Visit the Institute for Research on Unlimited Love
www.unlimitedloveinstitute.org

Explore the HeartMath Institute. Apply its recommendations.
www.heartmath.org

Learn about the Institute of Noetic Sciences (IONS)
www.noetic.org/research/projects/global-consciousness-project

3. See additional resources and recommended books, products and services at:

https://www.RAISEtheZONE.com
https://www.powerupbooks.com

4. Uphold Truth

Use NPR guidelines or another source to help identify fake news
www.npr.org/2016/12/11/505154631/a-finders-guide-to-facts

5. Generate Coherence in Your World

Practice meditation or physical activity that reduces stress and increases coherence in your mind and body, e.g.:

Transcendental Meditation™ - www.tm.org

HeartMath Heart-Focused Breathing
https://www.heartmath.org/resources/solutions-for-stress/

Mindfulness meditation, yoga, tai chi, dance, or recreation

Play music that reduces personal and societal stress. Play it in your home, school or business, or through a radio station
https://www.maharishiayurveda.uk/about-maharishi-gandharva-ved-music/

Listen to the Solfeggio frequencies
www.miraclesandinspiration.com/solfeggiofrequencies.html

6. Leverage Your Political Power

Get informed about the ballot initiative process in your city, county, and state. Learn from the following resources:

Read *Citizen Lawmakers* – by David Schmidt [includes a manual for organizing a campaign.]

7. Support solution-oriented candidates, or become one!

By completing this tour, you now have a great advantage in competing for a political office over candidates that have not been on this tour. Raise your zone. Raise your community!

Appendix B: Ballot Initiative Concepts

Following are sections of actual and proposed ballot initiatives aimed to achieve better public policy. They address key issues and offer cost-effective solutions. They are in chronological order.

2000 – State Budgeting [Submitted to the Colorado State Legislature. Legislative hearing held. Initiative withdrawn.]

- A proposed state law, "To require that a Perfect Budget Board…determine whether any proposed program for which a budget request is made, is efficacious or cost-effective according to peer-reviewed scientific journals, in achieving the goal of the requesting department, agency, or institution."

2000 – Perfect Health Board [Submitted to the Colorado State Legislature. Legislative hearing held. Initiative withdrawn.]

- A proposed state law, "…requiring insurance carriers to cover all health care services, including stress reduction methods, which are scientifically recognized to be efficacious and cost-effective."

2000 – Addressing Lying by Politicians [Drafted December 2000]

- All powers of every branch and every level of government are derived from the power of the People. The proposed law would require that when any elected or appointed government official, or their representative, addresses the court of public opinion on matters of public policy, they are addressing the highest court in the nation and are therefore under oath to tell the truth or be subject to prosecution and punishment for perjury.

2000 –Truth in Labeling for Fake News [Drafted December 2000]

- Information that is deceptive, fabricated, and blatantly false, is not only useless but also destructive with respect to accomplishing our goals as individuals and a society. This proposed ordinance would make it unlawful for any person to sell, give, deliver, furnish or broadcast any fake news without labeling it as false or fake.

2003 – Safety through Peace – Counterterrorism Initiative 101 [Placed on the November 2003 ballot in Denver, Colorado. See the full text at www.powerupbooks.com/resources for ballot initiatives]

- This proposed ordinance would have required the City and County of Denver to identify and implement any peace-creating programs, specifically programs that reduce societal stress as measured by scientific evidence showing that they greatly reduce crime, warfare and terrorism, especially such programs that also provide a net financial benefit to the city as a "peace dividend."

2010 – Creating an Extraterrestrial Affairs Commission – Initiative 300 [Placed on the November 2010 ballot in Denver, CO]. See the full text at www.powerupbooks.com/resources]

- This proposed ordinance would have required the City and County of Denver to, among other things, "provide the most accurate, complete, credible and relevant information about extraterrestrial intelligent beings on Earth to city government personnel and residents."

2013 – Protecting Extraordinary Technology [Filed with the State of Colorado* (2017), Ventura County, CA* (2018), and City of Denver, CO* (2019) [*Variations of the text below]

Article ___. PROTECTING EXTRAORDINARY TECHNOLOGY

Sec. ____. Definitions. As used in this article.

(1) *Extraordinary technology* means areas of innovation including, without limitation:

1. Industry forming/influencing inventions that are pioneering in scope
2. A motor or power plant, which has exceptionally high, non-polluting, non-hazardous, and sustained energy output relative to input
3. Inventions of demonstrated efficacy, (i.e. produces an intended result of value and usefulness), even if an invention appears to violate the predominant scientific understanding of physics or chemistry (antigravity, faster than the speed of light, accessing dark energy, etc.)

4. Room temperature superconductivity
5. Anti-global warming devices or any other device operating at the global scale
6. Methods or compositions for prolonging life or preventing aging
7. Prevention or curing of diseases previously considered impossible to prevent or cure, such as cancer, AIDS, HIV infection, dementia, Alzheimer's, COVID 19 and related variants, etc.)
8. Treatments to enhance intelligence

(2) Patent means a property right granted by the City and County of Denver to an inventor to exclude others from making, using, offering for sale, or selling the invention throughout the City and County of Denver or importing the invention into the City and County of Denver, for a limited time in exchange for limited public disclosure of the invention when the patent is granted.

Sec. ____. Declaration of findings and intent.

The People of the City and County of Denver hereby declare that the unfair and illegal suppression by the U.S. Patent and Trademark Office (USPTO) of cures for diseases, and inventions of cleaner energy and other extraordinary technologies, has resulted in unnecessary suffering and death to tens of millions of people and vast environmental destruction throughout the world....
[See full declaration at https://powerupbooks.com/resources]

The illegal secret USPTO Sensitive Application Warning System (SAWS), used during 1994-2015, delayed targeted patent applications indefinitely without explanation to patent applicants. **Targeted applications included: [Source: USPTO]**

- "Anti-Global Warming devices or any other device operating at the global scale"
- "Motor, Power plant, or other device which is self-sustaining (perpetual motion) or appears to violate the laws of chemistry or physics
- "Claiming prevention or curing of diseases which were previously considered impossible to prevent or cure."
- "Applications which would potentially generate unwanted media coverage (i.e., news, blogs, forums)"

Federal preemption of lower courts regarding federally issued patents does not prohibit any U.S. city, county or state from issuing patents, within their jurisdictions, for technologies that the USPTO has chosen to unfairly or illegally obstruct or delay.

Sec. ___. Protecting Extraordinary Technology

1. The City and County of Denver shall provide expedited issuance of patents for extraordinary technology, giving highest priority, without being a prerequisite, to categories of patent applications that were, or continue to be, unfairly or illegally obstructed or delayed by the U.S. Patent and Trademark Office. Funding for processing patent applications and issuing patents shall be obtained through application fees, or otherwise, as determined by the City and County of Denver.

2. The City and County of Denver shall affirm through a public proclamation and general policy, the legality of research, development, manufacturing, marketing, sale, possession, and use of extraordinary technology, regardless of patenting status.

Sec. ___. Implementation and Enforcement.

By the effective date, as set forth below, the City and County of Denver shall use all administrative rules, policies and procedures necessary for the proper and effective implementation and enforcement of the provisions of this article.

Sec. ___. Effective date.

This article shall be effective immediately from and after the date of its enactment.

Index

A *Finder's Guide to Facts* (NPR), 25
Aburdene, Ph.D., Patricia, 136, 137, 148
"Achilles' heel" of democracy, 82, 95
air pollution, 8, 9, 92, 159, 188, 189, 197 205
Allen, Ethan, 51
Alien Rock, 77
Altered Genes, Twisted Truth, 36
altruism, 137-141, 146-149, 173, 178
Amazing Mouse, 168, 171, 177, 182, 183
Amen, Dr. Daniel Gregory, 160
American Lung Association, 199
American Petroleum Institute, 89
Analemma, 189
Ancient Aliens, 52,59
Anniston, Alabama, 37
Archimedes, 131, 197, 201
Aristotle, 135, 142
artificial intelligence, 78
  hallucinations, 27, 28
ascension, 108
asthma, 84, 158, 159, 163, 202
Atlas Shrugged, 137
atomic bomb, 90, 91, 131
Atomic Energy Commission, 90
atomic fracking, 90

Baby steps, 85, 86
ballot initiative, 31-49, 63, 127, 215-218
Barela, Veronica, 72
Bassett, Stephen, 68
Better Than Cheese, 176, 183
biophotons, 79
Blake, William, 151, 152, 156
Blue Cloud, 6, 7
blue ribbon commission, 85, 86
Boyles, Peter, 94
Braden, Gregg, 180
Brain Fingerprinting, 118, 119
Bryant, Larry W., 76
bubble up economics, 194, 195
bullying, 2, 18, 160, 162
Burning Man Festival, 181
Butterfly Effect, 179, 193

Cancer, 24, 158, 159
capitalism, 136-144
capitalists, 141, *See* Hitler
Castle Bravo, 1
Castle, Odescalchi-Orsini, 11
CBS Evening News, 46
celebrities, 8-9
cell phone radiation, 159
Center for Responsive Politics, 91
chaos, 23, 96, 98, 161, 179, 185-187
  environment, 185
  global, 23
  theory, 179
Childre, Doc, 180
Chroma, Chuck, 72, 73, 77
Citizen Lawmakers, 34, 35, 214
civic organizations, 71
climate change, 7, 8, 95, 113, 197, 201
Clinton, President Bill, 35
coal reserves, 88
COCUUN, 186-206
coherence, 25, 40, 95-97, 117, 132, 136, 144, 148-168, 173-204, 207, 208
  brainwave, 96, 152, 164, 174, 189
  electrical, 196
  fuel plasma, 188, 189, 197, 204-206
  global, 175, 176, 178, 180, 184, 190, 191
  societal, 165, 166, 184
  water, 189, 207
COherent CUrrent UNion, 186
collective consciousness, 40, 82, 83, 95-97, 161, 183, 187, 189
collective stress, 47, 160
collectivism, 139, 140
Colorado Energy Office, 120
Colorado Greenhouse Gas Reduction Roadmap 2.0, 121
Columbine High School, 35
commitment, 9, 10, 116
communism, 138-140
consensus, 86
conservative radio monster, 47
constructive interference, 153, 202

Contact (movie), 15, 16
corrections industry, 94

D*aily Show with Jon Stewart,* 47
DaVinci Institute, 118
Deep State, 96
Democratic Party, 72-74, 111
Demon-Haunted World, 16
Denver City Council, 39, 42, 72
Denver Post, 35, 66, 72
devolved society, 96
direct democracy, 32, 33, 35
disease care industry, 94, 194
DNA, 100, 102, 190, 191
Donne, John, 145, 147
Druker, Steven J., 36
duopoly, 95-99

Earth Day, 94
Earthshot Prize, 204, 209
economy, 80-97, 125, 135-138, 192-196, 205
Edison, Thomas, 125
Einstein, Albert, 132, 143
Eisenstein, Charles, 192, 193
electric motors, 200, 201
electromagnetic fields (EMF), 184-186, 188, 190, 198, 201
electrostatic trap, 198
Emerson, Ralph Waldo, 50
enlightened self-interest, 136, 137, 143, 148
enlightenment, 163
entertainment industry, 187, 188
entrainment, 172, 180, 182
epigenetics, 158
expanded thinking, 50-53, 61, 78, 14, 131, 195
Extraterrestrial Affairs Commission, 64, 67, 72
extraterrestrial visitors, 54, 58, 59, 63, 64, 67-69, 74
Exxon, 90

Fagan, Dr. John, 36
fake news, 14, 21-24, 34, 215
Fantastic Reality, 69
Farish, William, 89, 90, 141, *See* Hitler
First Amendment, 32, 109
flood the zone, 114
fossil fuel industry, 132, 194
free market philosophy, 137
Franklin, Benjamin, 22, 51
Fuller, Buckminster, 5, 29, 81, 86, 106
Funes, Fr. José Gabriel, 57

Gandharva Veda Music, 191, 192
Gaynor, M.D., Mitchell L., 158, 159
genetic expression (and stress), 158, 159
Giving Xperiment, 168-171, 176-178, 183, 208
glass rod (physics), 157, 160
Global Carbon Reduction Grid, 201, 206
Global Coherence Grid, 190, 191, 196
Global Coherence Initiative, 175, 180, 207
Global Competitiveness Forum, 58
Global Consciousness Project, 181, 207
global transformation, 149, 193, 195
GMO, 36-39
graphene, 121, 131, 186
gratitude, 53, 170, 179
Greater Good Science Center, 173
greenhouse gas (GHG) emissions 121, 200, 206
Greer, Dr. Steven, 54

Hagelin, Dr. John, 47, 146, 162, 166
Haramein, Nassim, 207
Harris, Paola Leopizzi, 76
harmony, 133, 144-146, 150-156, 168, 174, 179, 191, 196
heart intelligence, 156, 180
HeartMath, 156, 174-176, 189, 207
Hitler, Adolf, 21, 89, 90, 141
Hopi Elders, 191
Horowitz, Dr. Leonard, 191
hydrogen bomb, 1
hydrogen, 120, 121, 205

Immune system, 158, 173
Initiative 101, 39, 41, 42, 46-48
Initiative 300, 64, 67-69, 72-76
Institute for Research on Unlimited Love, 170
Institute of Initiative and Referendum, 35
Institute of Noetic Sciences, 191, 207, 213
inventors, 91, 128, 129, 133

Jail (Denver), 45, 46, 87

Jefferson, Thomas, 33, 80, 114, 150, 161, 162
Jobs, Steve, 30, 86, 130
Juneteenth, 70

Kennedy, President John F., 50, 53
Kennedy, Robert, 139
Klein, Naomi, 89
know thyself, 162

Law of Attraction, 153
Letterman, David, 69
Los Angeles (air quality), 8, 93, 199
lying, 21, 27, 34

Madison, James, 29, 33, 106
Maharishi Effect, 166, 184
Maharishi International University, 157
Maharishi Mahesh Yogi, 82
Mass Electromagnetic Stress Syndrome, 159
McCartney, Sir Paul, 38
McConnell, John, 94
McCraty, Dr. Rollin, 170
Measure 27 - Oregon, 37-39
Megatrends 2010, 136
Meissner Effect, 164, 165
meliorism, 135
Mendocha, John, 83
M.E.S.S., 159, 207
Mindfulness, 180
Mirror image (of collective consciousness), 82, 83, 92, 122
Mitchell, Dr. Edgar, 145, 158, 181
Monsanto, 37
moonshot, 50, 177, 178, 204, 209
Murray, William H., 2, 9

NASA, 54, 58, 61, 76
national economic security, 125
National Press Club, 41, 54
National Public Radio, 25, 51
Native American, 6-8, 22
Nazi Germany, 90, 141
No Man is an Island, 145
noise (EMF), 159, 185-188, 198, 201
nuclear fusion, 91

Occam's razor, 15-17, 24, 138
One World Heart Project, 169, 214
Oppenheimer, Robert, 131
outdoor air purifiers, 202
ozone, 9, 199

Paradigm Research Group, 68
Parkinson, Robert G., 22
particle-wave duality, 142
Pentagon, 58, 61, 70, 76
pharmaceutical industry, 194
Pink UFO, 69
Plato, 3, 33, 105, 106
Polls (UFOs and Extraterrestrials), 65-67
Post, Ph.D., Stephen, 169
POTUS, 101, 102, 104, 107, 109
power grid, 190, 199-204
Power of the People, 31, 32, 128
PR Watch, 91
practical politics, 32
prefrontal cortex, 21, 160, 190
problem-based economy, 80, 87, 88, 92, 97, 125, 135
Project Rulison, 90

Quantum computing, 190

Radin, Dean, 172, 175
Rain Melody, 191, 192
RAISE the ZONE, ix, xi, 9, 59, 92, 114, 188, 189, 194, 195, 209
random number generators (RNG), 176, 181
Rather, Dan, 46
Redford, Robert, 9
Renewable and Sustainable Energy Review, 200
resonance, 170
revolving door (of government), 91, 92, 113
RF radiation, 159
Rik Veda, 146
Rockefeller Foundation, 88
Rocky Mountain Institute (RMI), 120, 121
Rosen, Mike, 47
Royal Society, 57
Rube Goldberg machine government, 96
Russell, Bertrand, 18

Sacred Economics, 192
Safety through Peace, 39, 41, 42, 44-48
Sagan, Carl, 16, 17, 79

Schopenhauer, Arthur, 45, 76
Science of Generosity Initiative, 173
secession, 107, 108
Secrecy Order, USPTO, 125, 133
self-government, 52, 96, 97
selfishness, 83, 137, 138, 140, 143, 145, 146
selfness, 137, 147, 148, 194
Sensitive Application Warning System (SAWS), 126, 133, 217
signal noise, 187
Singh, Major General Kulwant, 41
Small Modular Reactors, 91
Social Indicators Research, 166
societal stress, 40, 160, 185, 190, 192
Socrates, 25
Solfeggio frequencies, 214
solution-based civilization, 135, 136, 143, 148, 149, 151, 156, 192, 194
sphere of identity, 145, 146, 149, 194
Standard Oil of New Jersey, 89, 90, 141
suicide, 148, 160, 190
Summit on School Violence, 35
superconductor, 164
Sustainable Entertainment Alliance, 188
suppressed technology, 117, 122, 125
Swiss cheese (brains), 160
synchrony, 151, 152, 170

Templeton Foundation, 170
terrorism, 39-42, 48, 91, 98, 100-105, 108, 113, 118, 127, 146
terrorist, domestic, 41, 83, 91, 98, 101-104, 107-109
Tesla, Nikola, 29-31, 86, 125, 126, 132, 150, 153, 156, 202
Ticciati, Laura, 36
Titanic, 115, 195
Transcendental Meditation, 5, 163, 164, 166
triple filter test, 25
truth, 30-37, 45, 49, 52, 63, 69, 74, 75, 82-84, 96, 114, 147, 150-156, 163, 168, 193, 194, 196, 208, 209

U.C. Berkeley, 21, 141
U.S. Food and Drug Administration, 36
U.S. Patent and Trademark Office (USPTO), 125, 127, 128, 217
UNICEF, 198
unidentified aerial phenomena (UAP), 54
unified field, 47, 146, 151, 162
USA PATRIOT Act, 90, 102, 103, 107, 127

Vatican, 57, 76
violent crime, 42, 44, 48, 166
virus of incoherence, 96, 97, 160, 187, 188
Vuksinich, Cam, 169

Weapon of mass destruction, 23, 28
weapons industry, 94, 194
Wired Magazine, 125
wireless devices, 159
World Health Organization, 92, 198

Yoga Sutras, 164, 166
You Expect Success (Y.E.S.), 130, 131

Ending at the Beginning

This is the end of my book and our tour, but not the end of my journey or yours. Perhaps our paths will cross. I hope that the time and attention you gave to *RAISE the ZONE* will reward you in many ways for many years.

We've come full circle. The photo of me below was taken at my family's farm house that was shown in the beginning of this tour. I was already curious and pondering the nature of things.

My beloved parents always woke up my siblings and myself by urging us to "rise and shine!". That phrase has much more meaning to me now. I see a bright future for humanity as we "rise and shine" together to create Earth's Luminous Age.

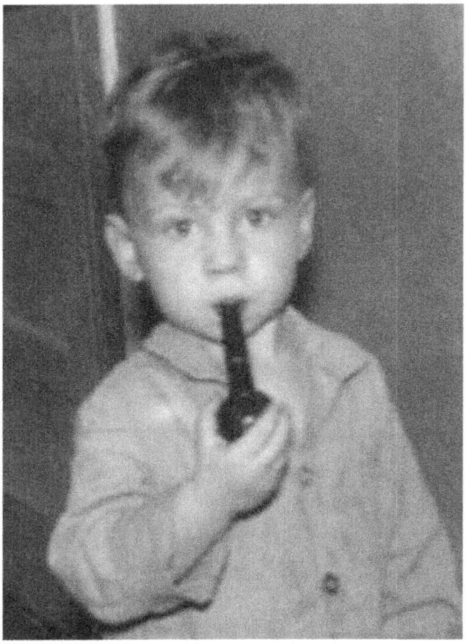

[Jeff Peckman at age four]

Made in the USA
Monee, IL
05 November 2026

67947702R00142